The Forecasting of Volcanic Eruptions

The Forecasting of Volcanic Eruptions

R. B. TROMBLEY, Ph. D.

iUniverse, Inc.
New York Lincoln Shanghai

The Forecasting of Volcanic Eruptions

iUniverse books may be ordered through booksellers or by contacting:

iUniverse
2021 Pine Lake Road, Suite 100
Lincoln, NE 68512
www.iuniverse.com
1-800-Authors (1-800-288-4677)

ISBN-13: 978-0-595-41260-0 (pbk)
ISBN-13: 978-0-595-85614-5 (ebk)
ISBN-10: 0-595-41260-2 (pbk)
ISBN-10: 0-595-85614-4 (ebk)

Printed in the United States of America

****

Dedicated to—

Mom & Dad, sister Karen

and all my former & current students.

****

TABLE OF CONTENTS

LIST OF FIGURES

LIST OF TABLES

PREFACE

It has been a long time for this writer to actually set down to write this document. But after eighteen years of development and the relative success of the results of the research, it was felt that now was the time to put it all together.

The author admits that initially in 1988 when the research was originally begun, it was met with a lot of criticism and skepticism. "Ridiculous...", "It can't be done...", "Impossible...", "You got to be nuts...", were but a few of the typical comments put forth when the project was started, as with many of us scientists who take on new projects, criticism it to be expected and is actually, in most cases, very helpful. Probably due to my scientific training and experiences, this author was not discouraged but rather encouraged by these comments. It is difficult for this writer's nature and make-up to accept the concept of *"It can not be done."*.

About six years after this writer started the project, a manuscript was prepared for submission to the coveted American Geophysical Union (AGU). After five revisions and re-submissions, it was finally accepted for publication in 1995. During those peer reviews, which are inevitable whenever an author submits a paper for publication, there were some excellent points and shortcomings brought up and subsequently worked into the paper prior to the paper's acceptance.

One of this book's aims is to inform the reader of the current algorithms, logic, mathematics, etc., that is currently being performed. It is also the aim to allow the reader, if so motivated, to develop their own software using the proposed algorithms and mathematics. As an alternative, may elect to use the already developed software. Particulars can be found in the Southwest Volcano Research Centre (SWVRC) website at http://www.swvrc.org.

The resulting work and now active software ***Eruption Pro 10.6*** is the cumulative product of all the research, studies, trials and scrutiny of paper submissions to many conferences. It has been a long struggle, a lot of work, but an extremely rewarding endeavour which to this writer, as with many other scientists, that dares to "...*go where no one has gone before.*"

—R. B. Trombley

CHAPTER 1
INTRODUCTION

Forecasting the time, place, and character of a volcanic eruption is one of the major goals of volcanology. It is also one of the most difficult goals to achieve. Until recently, people living in a volcano's shadow had little help anticipating an eruption. A major volcanic event might strike with no warning at all. In the past 300 years, volcanic eruptions, most of them unexpected, have killed more than 250,000 people. In 2000, experts estimated that 500 million people were living in areas at risk from catastrophic volcanic eruptions. This document describes the strides we have made in eruption forecasting in recent years and explores why accurately predicting volcanic events remains difficult.

The majority of volcanoes occur along the boundaries of tectonic plates. At these violent places, slabs of solid rock 5 to 100 kilometres (3 to 62 miles) thick are created and destroyed. Subduction zones, where oceanic crust is forced back into Earth's molten mantle, give rise to 80 percent of the world's volcanoes. The circum-Pacific belt, or Ring of Fire, which stretches up the western coasts of South and North America, across the Aleutian Islands, and down the eastern coasts of Asia, is one such zone.

Areas where two plates move away from each other, called rift zones, also give rise to volcanic activity. The East African Rift is a wide crack in Earth's crust that runs for more than 1,930 kilometres (1,200 miles) from Malawi through Tanzania, Kenya, and Ethiopia to the Gulf of Aden. Along this rift is a string of active volcanoes that includes Nyiragongo, a so-called stratovolcano located just north of the city of

Goma in the Democratic Republic of Congo. Nyiragongo is one of the world's most active and dangerous volcanoes.

In an effort to forecast volcanic eruptions, volcanologists monitor many types of activity within and around volcanoes: seismic activity, ground deformation, and gas emission. For example, to measure seismic activity within a volcano, scientists use a seismograph—the same apparatus used to measure earthquakes—to record vibrations of Earth's crust.

When rocks crack or slip past each other or when magma applies pressure to the inside of a volcano's magma channel, the surrounding rock vibrates much the way a tuning fork vibrates when it is struck. Seismographs record both the intensity and the frequency of the vibration, providing information that helps scientists determine the location, amount, and type of activity within a volcano.

Deformation of the ground on a volcano's slopes and gases emitted from a volcano's vents provide external signs of activity within the volcano. Pressure building inside a volcano's magma chamber or lava tubes may cause the ground surface to bulge outward. Satellite images and sensitive instruments positioned on a volcano's slopes allow scientists to measure elevation changes of a few centimetres or less. Analysis of the composition of gases emitted from a volcano also helps scientists determine the composition of magma inside, an indication of the volcano's explosiveness. Plus there are many other areas that are monitored at this time.

Unfortunately, no two volcanoes are alike. A combination that causes one volcano to erupt violently may cause no such result with another volcano. Having a limited number of cases to study, scientists are only beginning to understand the reasons for these variations.

Scientists can often find clues about past eruptions by studying the deposits left behind. Areas affected by lava flows, debris flows, tephra, or pyroclastic flows which can be mapped, making disaster planning more effective. In addition to this type of long-range forecasting, scientists are becoming more and more skilled at spotting the warning signs of an eruption.

A Potential Solution

An experimental computer programme, specifically designed for the MS-DOS & Windows based PCs has been developed and tested over the past eighteen years in an attempt to forecast long-range volcanic eruptions. The software package, *Eruption Pro 10.6* performs a statistical analysis on loaded volcano data from both historical and current real-time or near real-time data on all current active volcanoes about the world. It is the primary purpose of this work to provide the details, logic, algorithms and results to date of the aforementioned eighteen years of research, thought and experimentation that has been undertaken.

The *Eruption Pro 10.6* software package's intent is to forecast the next eruption event of volcanoes about the world. This software programme is intended as an additional forecast aid and diagnostic tool, and is not intended as the definitive concept in forecasting an eruption of any particular volcano. It should be kept in mind that the software package *Eruption Pro 10.6*, at this point, is in no way infallible and a forecast is only as good as the data used in creating it. The term "forecast" is used as it lends itself to a more probabilistic and less precise connotation of a precise scientific prediction, which has the connotation of precision. The current state-of-the-art in the discipline of volcanic forecasting is far from precise. Furthermore, forecasting as used by *Eruption Pro 10.6* has the notion of "*may* or *probably*" and not *will* erupt.

What is Eruption Pro 10.6?

Eruption Pro 10.6, to our knowledge, it is the only software programme of its type anywhere in the world. *Eruption Pro 10.6* performs analysis on current available volcano eruption data from both historical and current available eruption data, near real-time measurement data including, seismic, deformation, thermal, frequency of eruption analysis, solar & lunar influences, crater lake temperature (if applicable), COSPEC, & statistical procedures. The newest version also accounts for, albeit very small, contributions due to lunar and solar influences. It produces three forecasts; a statistically projected next eruption year, the next forecasted beginning eruption year with an ≥50% probability of eruption occurrence, and finally, the next forecasted beginning eruption year with an ≥ 95% probability of eruption occurrence. This software engineering package has been under development for over nineteen years—since 1987. *Eruption Pro 10.6* is a continually developing computer programme, specifically designed for the PC type computer in an MS-DOS version.

Eruption Pro 10.6 is available in the MS-DOS version and it is available in the English language only. The updates are automatically e-mailed within 24 hours to users and at no charge as events and/or updates occur. There is a great deal of information obtainable for each of the 498 volcanoes in the database. The new and updated version of *Eruption Pro 10.6*, was released on 15 August 2005.

This software programme is intended as an additional aid and diagnostic tool and is not intended as the definitive concept in forecasting an eruption of any particular volcano. It should be kept in mind that the software package *Eruption Pro 10.6* certainly, at this point, is in no way infallible, and only as good as the data used in creating it. This programme may now be used for volcanic hazard prediction or disaster mitigation by the public at this time. But it should be used only under the direction & guidance of scientists and those involved in volcanic hazard prediction or disaster mitigation.

References

1. Trombley, R. B., 1995, "*A Computer Based Long-Range Volcano Eruption Forecasting Programme, "Eruption"* ", EOS, Transactions, American Geophysical Union (AGU) Vol 76, No. 46, 7 Nov 1995 Fall Convention. Revised May, 1996.
2. Trombley, R. B., 1990, "*Computer modeling of statistical explosive patterns and the probability of volcanic event forecasting.*", Digital Equipment Corporation, U.S. Education Services, white paper.
3. Trombley, R. B., 2000, T. A. Jackson (Ed), *Caribbean Geology— Into The Third Millenium,* Chapter 23.

CHAPTER 2
THE BASIC PRECEPTS

In this Chapter we will explore the basic precepts of the concept of the forecasting of volcanic eruptions. The basic guidelines and understandings will be laid out and portrayed that will serve as the basis for many of the algorithms that are depicted within this book.

Defining Eruptions & Long-Range Forecasts

Whenever the discussion of volcanoes arises, the subject of eruptions is inevitable. But just what constitutes an eruption of a volcano becomes a valid point and is, of course, of concern and importance to input data to *Eruption Pro 10.6*.

In the 2nd Edition of *"Volcanoes Of The World"*, by Simkin and Siebert (1993), they define an eruption in the following manner, *"The arrival of volcanic products at the Earth's surface is termed an eruption."* Further, they go on to say, *"......we confine the term to events that involve the explosive ejection of fragmental material, the effusion of liquid lava, or both."* This is also the premise for *Eruption Pro 10.6* and only eruptions that produce pyroclastic materials, liquid lava, ash, phreatic eruptions or any combination of those volcanic products, are considered and entered into the database. Input data sources concerning the type of eruption, and relevant data are principally provided by three sources of data. Simkin and Siebert (1993), the account record as reported and published in the *"Volcanoes Of The World"*, the *"Bulletin of the Global Volcanism Network"* (Smithsonian Institution), and direct reports from actual visits and reports from various volcanic observatories and other responsible volcanic reporting agencies about the globe.

With respect to ***Eruption Pro 10.6's*** long-range forecasting ability, the term "long-range" used herein refers to the forecasting at least one (1) or more years in advance of an eruption event.

The Poisson Distribution Model

The Poisson distribution is a good model for describing phenomena where the probability of occurrence is small and constant. It arises as the model underlying various physical phenomena such as is the case with volcanic eruptions, which involve time. It is also an approximation where the number of trials, ***n***, is large as is the case of volcanoes where hundreds and even thousands of years pass before an eruption. The probability of success (an eruption), ***p***, is small. In other words, the Poisson distribution is an excellent distribution for rare events. As De La Cruz-Reyna (1991) states, "*If one concludes that well-sampled moderate-to-large magnitude sequences follow a Poisson distribution, then the basic features of Poissonian processes become fundamental in understanding the physics of volcanism. The analysis of published global data supports the notion that occurrence of eruptions can be accurately described as a simple Poisson process.*" The Poisson Distribution where μ = avg. eruption rate and x = # of successes is defined as:

$$P(x,\mu) = e^{-\mu} * \mu^x/x!$$

The Binomial Distribution Model

Shield volcanoes present a different diagnostic problem than do strato, complex, and compound volcanoes in that they do not follow a Poisson distribution. But shield volcanoes are similar to other types of volcanoes in that they either are erupting or not erupting. It appears that a Binomial distribution might be the best distribution fit for shield volcanoes.

For shield volcanoes, we consider a set of ***n*** mutually independent trials each made under these conditions and ask for the probability of exactly ***r*** successes (eruptions) and ***n − r*** failures (no eruptions). Each of these independent trials is, of course, a binomial distribution. Each trial is independent so the probability of a *specific* sequence, e.g., starting of with **r** successes followed by **n − r** failures is $\mathbf{p^r q^{n-r}}$. However, the order of the sequence is irrelevant. Any order of eruption (or non-eruption) events will do, and each possible order has the same probability of occurring, $\mathbf{p^r q^{n-r}}$.

We must, therefore, multiply this probability by the number of ways **n** trials can be divided into **r** successes (eruptions) and **n − r** failures (no eruption). This number is $\mathbf{_nC_r}$, and the overall probability required is

$$\mathbf{P_{n,p}(r) = \frac{n!}{r!\,(n-r)!}\ p^r q^{n-r}}$$

where

P(r) = Probability of an eruption

p = Probability of success (eruption) on any one trial

q = Probability of failure (no eruption) on any one trial

Hypergeometric Distribution:

Hypergeometric probability is determined to be the best probability distribution to be used with solar influences as the hypergeometric distribution is used to determine the probability of obtaining a certain sample outcome when sampling without replacement. The hypergeometric probability function is described as

$$\mathbf{f(x)} = \frac{\binom{r}{x}\binom{N-r}{n-x}}{\binom{N}{n}}$$

where r = number of elements in the population labeled successes, x = # of successes, n = number of trials, N = number of elements in the population, and f(x) = the probability of x successes in n trials.

Revised Probabilities

Revising probabilities when new information is obtained is an important part of probability analysis. Often, as is the case with most volcanoes assumed to be Poisson distributed, the initial or *prior* probability estimates are completed for a specific event of interest, i.e., the probability of an eruption for the current year. Then, some new additional information is obtained, a missed eruption, or the fact that another year transpires and there has been no eruptive event. Given this new information, the prior probabilities are updated by calculating the revised probabilities referred to as *posterior* probabilities. Bayes' Theorem provides a means for making such calculations. This theorem, along with the axioms suggested by the combining of Poisson distribution and negative binomial distribution and using a Bayesian analysis, as they apply to volcanic eruptions (Ho, 1990), have been incorporated into *Eruption Pro 10.6*.

When the Poisson process, as applied to volcanic eruptions, is expanded to accommodate a gamma mixing distribution on λ, there becomes an immediate consequence of this mixed Poisson model. The frequency distribution of eruptions in any given interval of equal time length follows a negative binomial distribution. The probability of x eruptions becomes:

$$P(x) = \frac{\Gamma(r + x)}{\Gamma(r)\, x!} \left[\alpha / (\alpha + 1)\right]^r \left[1/(\alpha + 1)\right]^x, \quad x = 0, 1, 2, \ldots\ldots$$

where r and α are the shape and scale parameters of the gamma distribution respectively. Treating the average eruption rate λ as a random variable means that the probability distribution function $f(x,\lambda)$ is, in reality, a *conditional* probability. The condition being that λ is in state

λ. Therefore, when using a probability distribution for λ, it is more suitable to use the notation $f(x|\lambda)$ for the data x. From the conditional distribution of x and the given (calculated) *prior* distribution for λ, the joint distribution of (x,λ) can be calculated. Thus:

$$f(x,\lambda) = f(x|\lambda)g(\lambda)$$

where $g(\lambda)$ is the probability density function and the marginal or absolute distribution of x, with probability:

$$P(x) = E_g[f(x,\lambda)] = \int f(x|\lambda)g(\lambda)\ d\lambda$$

For the volcanoes being monitored by *Eruption Pro 10.6*, and assuming that λ follows a gamma distribution, then

$$g(\lambda) = \frac{\underline{\alpha}^r\ \lambda^{r-1}\ e^{-\alpha\lambda}}{\Gamma(r)}\ ;\ 1 > 0;\ r,\alpha > 0$$

where r and α are the shape and scale parameters respectively as previously mentioned, and

$$f(x|\lambda) = \frac{e^{-\lambda}\ \lambda^x}{x!},\qquad x = 0, 1, ...$$

Therefore, from the $g(\lambda)$ equation above, the absolute probability for the number of eruptions per unit of time interval is given by,

$$P(x) = \int_0^\infty \frac{e^{-\lambda}\underline{\lambda}^x}{x!}\ \frac{\underline{\alpha}^r}{\Gamma(r)}\ \lambda^{r-1}\ e^{-\alpha\lambda}\ d\lambda$$

$$= \frac{\Gamma(r+x)}{\Gamma(r)\ x!}\ [\alpha/(\alpha+1)]^r\ [1/(\alpha+1)]^x\ ,\ x = 0, 1, 2, ...$$

The mean and variance for the negative binomial distribution are given by:

$$E(x) \ = \ r/\alpha$$

and $$\mathbf{Variance(x)} \ = \ r(\alpha + 1)/ \alpha^2 \ .$$

The incorporation of the combined negative binomial and Poisson distributions along with the Bayesian analysis has had a positive effect on the statistical forecast accuracy of *Eruption Pro 10.6.* The increased performance can be observed from the results of essentially two factors; a) the incorporation of the Bayesian analysis and b) the updated volcano eruption data incorporated into the software. These factors alone appear to have improved the forecasting ability of *Eruption Pro 10.6.*

Probability Contributions

In addition to the normal probability contribution in *Eruption Pro 10.6* from the historical data, there are several other contributions that contribute to the overall analysis. Those other contributions are: Input from Correlation Spectrometer (COSPEC), Crater Lake Thermal, Satellite Thermal Imaging, Volcanic-Seismicity, Deformation, Lunar & Solar Influences and the volcano's Frequency of Eruption analysis. The following discusses their input and how the contribution is used in *Eruption Pro 10.6.* The next Chapter begins with the seismicity contribution.

References:

1. De La Cruz-Reyna, S., 1991. *"Poisson-distributed patterns of explosive eruptive activity."*, Bulletin of Volcanology., 54:57-67.
2. Ho, Chih-Hsiang, 1990, *"Bayesian analysis of volcanic eruptions"*, Journal of Volcanology and Geothermal Research, 43: 91-98
3. Meyer, S. L., *"Data Analysis For Scientists And Engineers"*, 1975, Chapter 23, pgs.193-201

4. Simkin,T., Siebert, L., McClelland, L., Bridge, D., Newhall. C., Latter, J.H., *"Volcanoes of the World"*, Washington, D.C. : Smithsonian Institution, 1981.

5. Simkin, T., Siebert, L., McCleland, L., May, *"Volcanoes of the World"*, 1984 Supplement, Washington, D.C.: Smithsonian Institution, 1984.

6. Simkin, T., Siebert, L., *"Volcanoes of the World"*, 2nd Edition, Washington, D.C.: Smithsonian Institution, 1993.

7. Wickman, F. E., 1966, *"Repose patterns of volcanoes"*. *I. Volcanic eruptions regarded as random phenomena."*, Ark:Mineral. Geol.,4:291-301.

Chapter 3
Probability Contribution Due
To Seismic Analysis

Introduction

Seismicity plays an important and major role in the probability determination relative to a volcano's potential (probability) of having an eruptive event. It is, arguably the largest contributor to the probability calculations. At an active volcano, long-period seismicity (with typical periods in the range 0.2-2 s) reflects pressure fluctuations resulting from unsteady mass transport in the sub-surface plumbing system, and hence provides a glimpse of the internal dynamics of the volcanic edifice. When this activity occurs at shallow depths, it may signal the pressure-induced disruption of the steam-dominated region of the volcano, and can accordingly be a useful indicator of an impending eruption.

Volcanoes are the source of a great variety of seismic signals which behave differently than events on earthquake faults. Nearly every recorded volcanic eruption has been preceded by an increase in earthquake activity beneath or near the volcano, and accompanied and followed by varying levels of seismicity. For this reason, seismic analysis has become one of the most useful and important tools for eruption forecasting and monitoring. At the present time, approximately 200-300 of the world's volcanoes are seismically monitored, although the number and quality of stations at each varies considerably. This represents about one out of three of the approximately 547 volcanoes that have erupted in historic times. Over the last several decades, 40 to 60 individual volcanoes, on the average, have erupted each year. Because erupting volcanoes draw attention both publicly and scientifically, over half these are seismically monitored.

Types of Volcanic Earthquakes

There are earthquakes and then there are earthquakes specific to volcanoes. In recent years, we scientists have learned a great deal about the earthquakes volcanoes generate. Some volcano earthquakes originate in solid rock and some in molten rock that is trying to move through underground channels. It is the latter type of earthquake that helps forecast eruptions and shows promise in forecasting future eruptions.

Volcano Tectonic Events (VT)

Pressure from a pool of magma has just cracked solid rock, creating a volcano-tectonic (VT) event. This type of quake produces relatively high-frequency shaking, usually between one and five cycles per second. An increase in VT activity is often an early sign that a volcano is becoming active. This type of restlessness, however, can last anywhere from days to years, so it's not a reliable way by itself to predict when a volcano might erupt.

A VT event occurs when magma under pressure or cooling rock causes rock to crack or slip. The abrupt motion of the rock causes its seismic signal to appear abruptly on a seismogram. Even though the way they are produced is different, seismograms produced by volcano-tectonic earthquakes look like those produced by typical earthquakes (those caused by the motion of tectonic plates at plate boundaries, such as the San Andreas fault and the Mid-Atlantic Ridge).

VT events cycle as many as five times a second, particularly if the earthquake is two kilometres (1.2 miles) or more below the surface. The frequency of the VT signal shown in Figure 3-1 is five cycles per second.

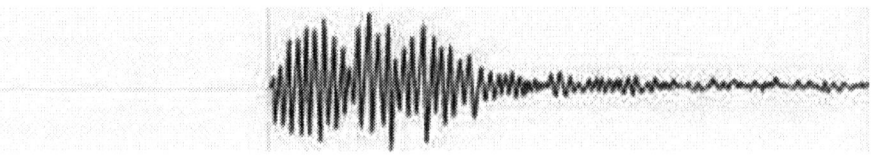

Figure 3-1. Volcano-Tectonic Event (VT) Signature

Long Period Events (LP)

Sudden changes in pressure within magma-filled cracks and channels cause long-period (LP) events. LP events are volcano-related earthquakes that are lower in frequency than volcano-tectonic (VT) events. The frequency of LP events is one half of a cycle to three cycles per second.

Unlike VT events, LP events can reveal magma flow and the buildup of pressure within a volcano. This knowledge can help us forecast eruptions.

The shaking that causes LP events is similar to the "water hammer" that happens in household water pipes. When water is moving quickly through a pipe and the faucet is turned off, the water is forced to stop. But instead of coming to an abrupt stop, it bounces against the closed valve, creating a wave of pressure that moves back and forth within the pipe. The rate at which the wave bounces is determined by the pipe's resonant frequency, a natural frequency of vibration that is, in turn, determined by several factors, including its length and shape. This bounce causes the pipe to clang loudly.

The same thing happens within a volcano's magma channel, except that the channel's end is already closed, and the abrupt change is caused by variations in the magma's pressure. Also, the frequency of the bounce is much slower within the channel. Figure 3-2 illustrates the signature of an LP event.

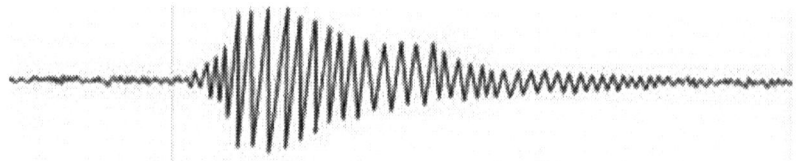

Figure 3-2. Long Period Event (LP) Signature

Tremor (LP)

A tremor is a long-period (LP) event, but one that lasts longer than the typical LP event. In fact, a single tremor can last anywhere from several minutes to months.

The frequency range of a tremor is the same as that with an LP event: one half of a cycle to three cycles per second. The signal shown in Figure 3-3 has a frequency of two cycles per second. Like LP events, tremors can also be a good indicator of an impending volcanic eruption.

The source of a tremor can and often is the same crack or channel that produces LP events. The difference is that, with a tremor, the waves of pressure traveling through the magma get a little extra push every so often. This push can be pressure changes coming through magma channels from below. Because the waves creating the tremor travel at the cracks' resonant frequency (see LP section above), the signal can appear as a continuous wave moving at a single frequency.

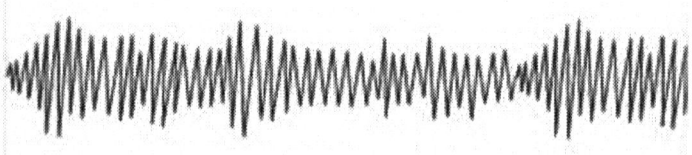

Figure 3-3. Tremor Event (LP) Signature

Hybrid Events (HY)

Sometimes a volcano-tectonic (VT) event triggers a long-period (LP) event, and vice versa. A seismic signal that contains a mixture of both types is called a hybrid (HY) event.

The hybrid event in Figure 3-4 is a VT event that triggered an LP event. Notice how the signal is bunched up more at the beginning than it is later on. The first part of this signal shows the VT event; later, the less-bunched (lower frequency) signal of the LP event appears.

Because LP events often begin with signals that look very similar to those at the beginning of a hybrid, it is usually difficult for volcanologists to distinguish between the two.

To tell one from the other, volcanologists look closely at several seismograms of a single event, as recorded by seismographs placed at various locations. If the first part of the signal looks similar on all of the seismograms, they probably have an LP event.

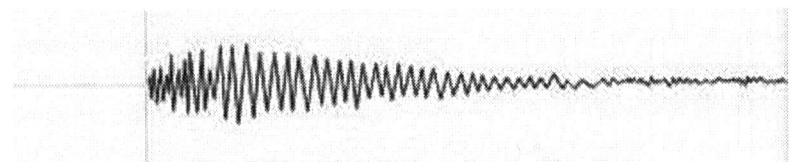

Figure 3-4. Hybrid Event (HY) Signature

Algorithms

Table 3-1 below contains values of the number of Long Period (LP), Volcano-Tectonic (VT), and Hybrid (HY) values. Based on these values received, real-time values, or combinations therein from reporting geological agencies about the world, a scaled value, ranging from 0 to .5

probability contribution for a particular volcano, is determined and entered into the ***Eruption Pro 10.6's*** software package's database.

As an example, if there were 7 VT events recorded at a particular volcano and 152 LP events recorded for the same volcano, then the probability contribution input would be .22 (highlighted on Table). This would serve as only one input to ***Eruption Pro 10.6's*** multitude of inputs prior to determining the probability of an eruptive event for the volcano examined.

As was previously mentioned, approximately 200-300 of the world's volcanoes are seismically monitored, although the number and quality of stations at each varies considerably. This implies that some volcanoes have no seismic signals to monitor and therefore report. This is, of course, the case when there is no seismic equipment available for the volcano or the volcano is too remote to be economically and successfully monitored. In the case where there is no input from an active volcano, ***Eruption Pro 10.6*** accounts for no VT, LP or HY counts by defaulting to a seismic probability contribution of .1. Since it is a well known fact that all volcanoes will exhibit some form and type of seismic activity prior to an actual eruption, a conservative value of .1, as the seismic contribution, was selected, albeit probably low, so as not to unfavourably bias the software and not take into account fully the other input algorithms.

TABLE 3-1. Seismic Probability Contributions

Factor	VT	Hybrids	LP
0.00	0	HY<10	LP<10
0.10	0<VT≤5	10<HY≤40	10<LP≤50
0.11		40<HY≤80	50<LP≤100
0.12		80<HY≤120	150<LP≤200
0.13		120<HY≤160	200<LP≤250
0.14		160<HY≤200	250<LP≤300
0.15		200<HY≤240	300<LP≤350
0.16		240<HY≤280	350<LP≤400
0.17		280<HY≤320	400<LP≤450
0.18		320<HY≤360	450<LP≤500
0.19		HY>360	LP>500
0.20	5<VT≤30	10<HY≤40	10<LP≤50
0.21		40<HY≤80	50<LP≤100
0.22		80<HY≤120	150<LP≤200
0.23		120<HY≤160	200<LP≤250
0.24		160<HY≤200	250<LP≤300
0.25		200<HY≤240	300<LP≤350
0.26		240<HY≤280	350<LP≤400
0.27		280<HY≤320	400<LP≤450
0.28		320<HY≤360	450<LP≤500
0.29		HY>360	LP>500

NOTE: VT, HY and LP all counts per day.

0 = Normal background seismicity.
.1 = Very light seismic activity.
.2 = Light seismic activity.
.3 = Moderate seismic activity.
.4 = Large seismic activity.
.5 = Very large & frequent seismic activity.

TABLE 3-1 (Cont'd). Seismic Probability Contributions

Factor	VT	Hybrids	LP
0.30	30<VT≤50	10<HY≤40	10<LP≤50
0.31		40<HY≤80	50<LP≤100
0.32		80<HY≤120	150<LP≤200
0.33		120<HY≤160	200<LP≤250
0.34		160<HY≤200	250<LP≤300
0.35		200<HY≤240	300<LP≤350
0.36		240<HY≤280	350<LP≤400
0.37		280<HY≤320	400<LP≤450
0.38		320<HY≤360	450<LP≤500
0.39		HY>360	LP>500
0.40	40<VT≤75	10<HY≤40	10<LP≤50
0.41		40<HY≤80	50<LP≤100
0.42		80<HY≤120	150<LP≤200
0.43		120<HY≤160	200<LP≤250
0.44		160<HY≤200	250<LP≤300
0.45		200<HY≤240	300<LP≤350
0.46		240<HY≤280	350<LP≤400
0.47		280<HY≤320	400<LP≤450
0.48		320<HY≤360	450<LP≤500
0.49		HY>360	LP>500
0.50	VT>75	HY>360	LP>500

NOTE: VT, HY and LP all counts per day.

 0 = Normal background seismicity.
.1 = Very light seismic activity.
.2 = Light seismic activity.
.3 = Moderate seismic activity.
.4 = Large seismic activity.
.5 = Very large & frequent seismic activity.

References

1. H. Sigurdsson, et al, 2000, *"Encyclopedia of Volcanoes"*, Volcanic Seismicity, Pg. 1016

2. R. B. Trombley, J. P. Toutain, 2002, *"ERUPTION Pro 10.5—The New & Improved Long-Range Eruption Forecasting Software"*, 16th Caribbean Geological Conference, Barbados, June, 2002, and published in the Transactions of the 16th Caribbean Geological Conference.

CHAPTER 4
PROBABILITY CONTRIBUTION DUE TO DEFORMATION ANALYSIS

Introduction

As with other contributions, ground deformation plays a role in the probability determination of an eruption. Ground deformation on volcanoes may be due to several causes. These include: (a) inflation/deflation of a buried magma storage zone, (b) injection of a dike or sil which may or may not be an eruption conduit, (c) subsidence due to lava loading, gravitational settling, or spreading of the entire volcano, and (d) slope movement caused by slope creep prior to failure or by magma pressure variations on steep slopes. Combinations of these causes frequently occur to produce a complex pattern of deformation. *Eruption Pro 10.6* will attempt to account for some of these contributions due to deformation.

Deformation Measurement Types

InSAR

Volcanoes deform constantly. Obviously during eruptions, the magma injected from down below has to work its way through surrounding rock, and this creates slow ground movements that can be detected with InSAR. But also when a volcano is in relative quiescence, we can monitor the injection of magma at depth because that deformation too reaches the surface. Again, this tool allows us to understand better how volcano work. InSAR stands for Interferometric Synthetic Aperture Radar. This is thus a remote sensing technique that uses radar satellite images. Those radar satellite (ERS1, ERS2, JERS, IRS or Radarsat) shoot constantly beams of radar waves towards the earth and record them after they bounced back off the Earth's surface.

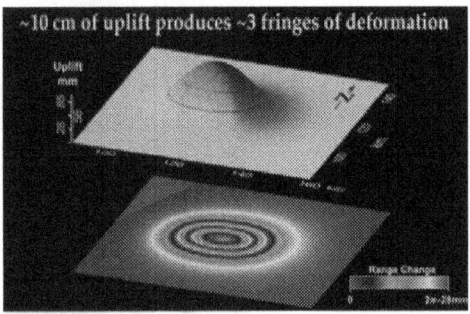

Figure 4-1. Example of an InSAR measurement.

Mogi Model

Another technique is to utilise the results of the Mogi model (Mogi, 1958) in its analysis. In active volcanic environments, a magma chamber inflation/deflation process model is usually applied to the data. The "Mogi" point source model, which is based on an elastically expanded point source in a half space (called a "Mogi point source"), usually serves to explain the observed deformation in such cases. In the case of Nisyros, the "Mogi model" was applied in order to explain the observed deformation deduced by DGPS measurements . The Mogi model predicts that

$$\Delta d = \frac{3a^3 Pd}{4\mu(f^2 + d^2)^{(3/2)}}$$

and

$$\Delta h = \frac{3a^3 Pf}{4\mu(f^2 + d^2)^{(3/2}}$$

where
  a = the radius of the source sphere
  P = change in hydrostatic pressure in the sphere
  f = depth to the centre of the sphere
  μ = Lame's constant
  d = radial distance on the surface from a point above the source
  Δd = radial horizontal displacement at the surface
  Δh = vertical displacement of a point at the surface

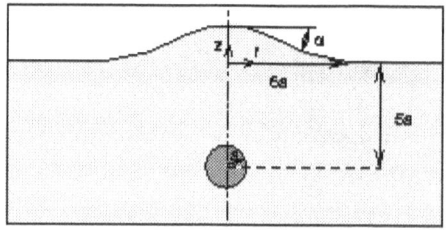

Figure 4-2. Example of an Mogi Model measurement.

GPS

The Global Positioning system consists of a constellation of 24 satellites. Each satellite orbits Earth twice a day at an altitude of about 20,000 km and continuously transmits information on specific radio frequencies to ground-based receivers. GPS was developed by U.S. Department of Defense as a worldwide navigation system and has been adopted by civilians for many other uses, including surveying, mapping and scientific applications. Relatively inexpensive GPS receivers like those used by pilots, boaters and outdoor enthusiasts can determine its position on the Earth's surface to within a few tens of metres. With more sophisticated receivers and data-analysis techniques, we can determine receiver positions to less than a centimetre. The Global Positioning System (GPS) can be utilised to detect ground deformations of the surface of a volcano. Ground deformation monitoring is considered one of the most effective tools for investigating the behaviour of active volcanoes. The decreasing cost of GPS hardware, together with the increased reliability of the technology, facilitates such demanding applications. GPS ground deformation measurements can be continuous, automatic, conducted in all weather conditions and provide useful data.

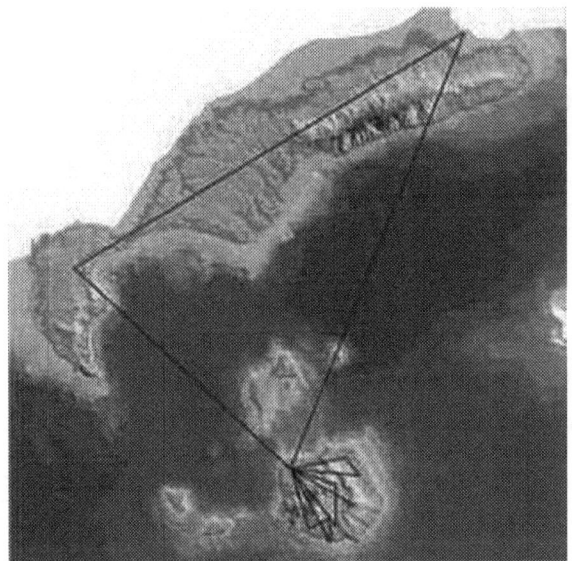

Figure 4-3. Example of a GPS layout.

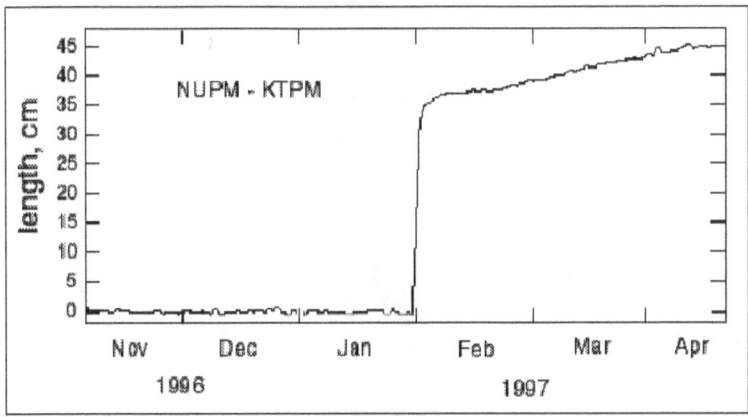

Figure 4-4. Example of a GPS measurement.

Tiltmeter

As the magma reservoir becomes inflated, the ground around it cracks to accommodate its increasing volume. Many small earthquakes occur in the area surrounding the magma as the rocks break. As the surface of the volcano changes shape, tiltmeters record tiny changes in slope, distances increase between benchmarks on opposite sides of the caldera, and elevations of the stations increase. In order to correctly interpret tiltmeter data, it is important to note the location of the tiltmeter in relation to the magma reservoir. For example, on volcano Mauna Loa, in Hawaii, tiltmeter station MOK is northwest of the summit magma chamber. A tilt to the northwest recorded at this station is, therefore, a tilt away from the magma chamber and is usually indicative of magma reservoir inflation. However, if the same tilt were recorded at a tiltmeter located on the southeast side of the magma chamber, this would be a tilt toward the magma chamber, indicating a deflation of the magma reservoir.

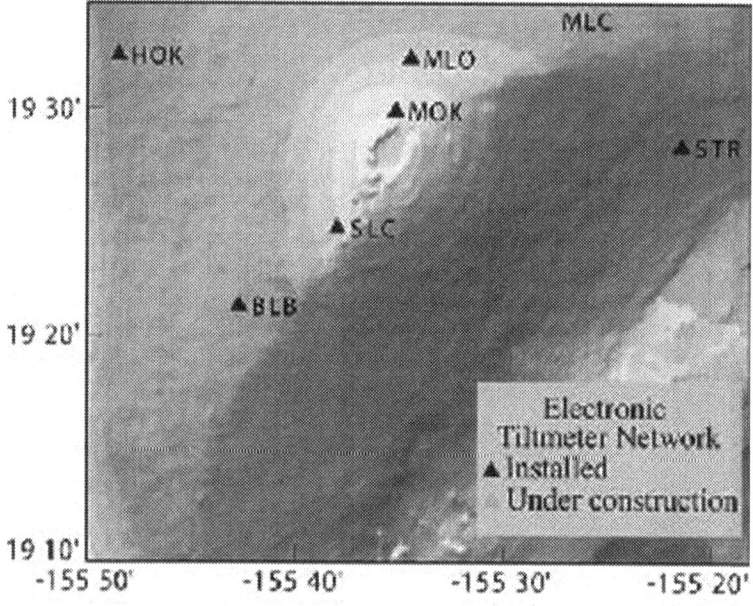

Figure 4-5. Example of a Tiltmeter layout.

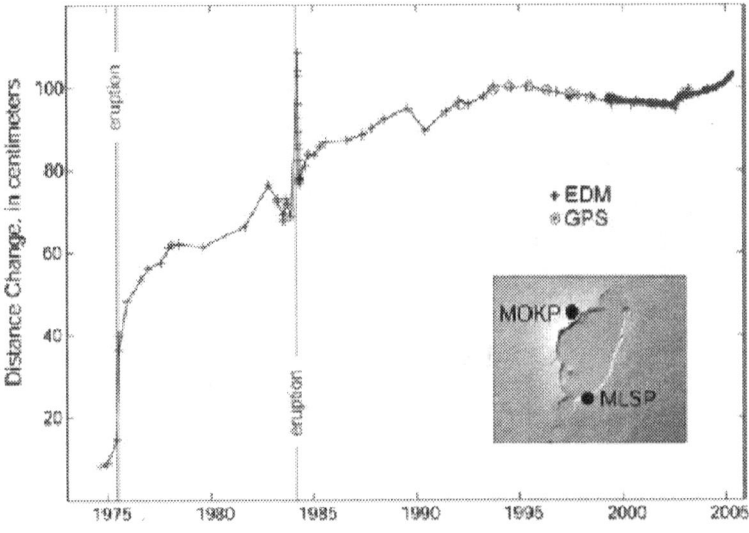

Figure 4-6. Example of a Tiltmeter measurement.

Algorithms

The input to the software, **Eruption Pro 10.6**, is in the form of a sliding scale function based on the received displacement with a minimum of 0 and a maximum of .2 As the deformation reading changes, so does the probability contribution. The schema currently used is as follows:

Technique	Reading	Probability Contribution
InSAR	<100 cm	0.0
	100 ≤ x <500 cm	0.5
	500 ≤ x <700 cm	1.0
	700 ≤ x <1000 cm	1.5
	>1000 cm	2.0
Technique	Reading	Probability Contribution
Mogi Model	The Mogi Model is not used by **Eruption Pro 10.6**.	

Technique	Reading	Probability Contribution	
GPS	<100 cm	0.0	
	100 ≤ x <500 cm	0.5	
	500 ≤ x <700 cm	1.0	
	700 ≤ x <1000 cm	1.5	
	>1000 cm	2.0	
Technique	Reading	Probability Contribution	
Tiltmeter	<.05 mr	0.0	(mr = microradians)
	.05 ≤ x <1 mr	0.5	
	1 ≤ x <2.0 mr	1.0	
	2.0 ≤ x <2.5 mr	1.5	
	>2.5 mr	2.0	

References:

1. Mogi, K., (1958) *"Relation between eruptions of various volcanoes and the deformation of the ground surfaces around them"*, Bull. Earthquake Res. Inst. Univ., Tokyo, 36, 99-134

2. R. B. Trombley, J. P. Toutain, 2002, *"ERUPTION Pro 10.5—The New & Improved Long-Range Eruption Forecasting* Software", 16th Caribbean Geological Conference, Barbados, June, 2002, and published in the Transactions of the 16th Caribbean Geological Conference.

3. GEOWARN,Geospatial warning systems URL:
   http://www.geowarn.ethz.ch/index.asp?ID=9&cat=121/

4. U.S. Department of the Interior, U.S. Geological Survey, Menlo Park, California, USA URL http://volcanoes.usgs.gov/About/What/Monitor/Deformation/GPS.html

5. U.S. Geological Survey, Earthquake Hazards Program URL: http://quake.wr.usgs.gov/research/deformation/modeling/

6. U.S. Geological Survey, Hawaii Volcano Observatory URL: http://hvo.wr.usgs.gov/maunaloa/current/main.html

CHAPTER 5
PROBABILITY CONTRIBUTION DUE TO
COSPEC ANALYSIS

Introduction

Gas studies include frequent airborne measurements of sulfur dioxide and carbon dioxide in plumes and less frequent sampling of gases from crater fumaroles. The emissions of sulfur dioxide are measured in the plume by a Correlation Spectrometer (COSPEC) designed originally for pollution studies. The instrument measures the amount of solar ultraviolet light absorbed by sulfur dioxide in the plume and compares it with an internal standard. Three to six traverses are made beneath the plume at right angles to the plume trajectory several times each week at a volcano to calculate daily emission rates.

Many active volcanoes release gases to the atmosphere both during and between eruptions. The main gas species emitted are H_2O, CO_2, H_2S, SO_2, H_2, CO, CH_4, HCl and HF, the relative proportions of which can be related to thermodynamic (temperature-pressure-oxygen) conditions. The COSPEC is a portable spectrometer which measures the absorption of solar ultraviolet light by means of SO_2 molecules. The *Eruption Pro 10.6* will attempt to account for some of these contributions due to the COSPEC analysis when data is obtained.

SO2 Flux Data

The SO_2 flux data currently supplied by COSPEC measurements are commonly used 1) to constrain the masses of magma that is degassing and 2) to correlate with the level of activity and therefore are suitable data for long time monitoring.

SO$_2$ Emission and Volcanic Activity

Volcanoes emit measurable SO_2 fluxes in conditions of low explosivity, effusive activity, dome or intrusion degassing or open-vent degassing. Typical SO_2 fluxes measured at 17 volcanoes showing different state of activity between 1984 and 1991. This suggests a classification of SO_2 emitters, with small (< 200 t/d), moderate (200-1000 t/d) and large (> 1000 t/d) emitters. Moderate and large SO_2 fluxes are considered as coming from magma degassing.

Long Time-Series

As most of SO_2 flux data are sporadic measurements performed over more or less short periods. It is interesting to observe really long-time continuous monitoring, such as those performed at Galeras (Columbia) from 1989 to 1995 or Soufriere Hills (Montserrat, West Indies) in 1997. At Galeras, low SO_2 fluxes were recorded after the May 1989 eruptions, indicating the presence of a shallow and partially degassed magma or a conduit that was partially closed. On the contrary, the very large SO_2 fluxes from September 1989 to March 1990 indicated that the magma was undegassed and the conduit was open.

Recent measurements have demonstrated that SO_2 fluxes were correlated with deformation rates and bulk volcanic-seismicity. It ahs been shown that at extrusive domes-type volcanoes, SO_2 emission rates were supposed to fluctuate as the result of various processes operating (release of gas through the dome and conduit, flow-retardation in free-spaces in dome, direct release, dome cracking by extrusion of magma, dome disruption by pyroclastic flows. This leads on this type of volcanoes to potentially very variable fluxes. At Soufriere Hills, SO_2 eruption rates are highly correlated with ground deformation in periods of high hybrid (mixed VT and LP events) volcanic-seismicity. At this volcano,

SO_2 flux, tilt amplitudes and hybrid volcanic-seismicity clearly increased during the 4 days prior the 25 June 1997 dome collapse.

SO_2 Conclusion

Many SO_2 flux measurements have been performed at active volcanoes. Evidence of the relationship between emission rates and activity are available in the literature. At most volcanoes, high variations of the SO_2 emission rate are recorded prior, during and after eruptions. Because exsolution of volatiles is controlled by many factors (permeability, pressure, viscosity, and porosity of domes) changing prior eruptions can be either increasing or decreasing rates, according to the type of volcanoes, the feeding system, and the dynamics of the volcano. Within this frame, volcanoes working with active domes should be considered as the most complex systems to monitor with SO_2 flux measurements.

Algorithms

COPEC readings are obtained from the various observatories and other official volcanic reporting agencies throughout the world as the readings are made and become available. ***Eruption Pro 10.6*** compares the nominal readings with the actual readings taken from the volcano under analysis. The comparison is performed from a ratio format from which the probability contribution is determined. e. g., volcano Soufriere Hills on the island of Montserrat has a nominal COSPEC reading of 450 tonnes per day output. The current actual reading is 640 tonnes per day. Therefore the ratio is calculated as:

$$\mathbf{R}_{COSPEC} = \frac{\mathbf{COSPEC}_{nominal}}{\mathbf{COPEC}_{actual}} = \frac{\mathbf{450}}{\mathbf{640}} = \mathbf{.142}$$

As the COSPEC reading waxes and wanes, so does the ratio and therefore the probability contribution. The software programme is safety interlocked at a maximum of .300 and a minimum of 0 as a probability contribution due to COSPEC.

As an alternative, if the SO_2 emissions with very small (<100 t/d), small (<200 t/d), moderate (200-1000 t/d) and large (>1000 t/d) emitters would have the following probability contributions assigned;

No Reading or (< 50 t/d)	**=**	**0**
Very Small (< 100 t/d)	**=**	**.05**
Small (< 200 t/d)	**=**	**.1**
Moderate (200-1000 t/d)	**=**	**.2**
Large (> 1000 t/d)	**=**	**.3**

References

1. Caroll, M.R. and Webster J.D. (1994) In Volatiles in magmas, Caroll & Holloway Eds., 231.
2. Doukas, A (1995) *"Compilation of Sulfur Dioxide and Carbon Dioxide Emission- Rate Data from Cook Inlet Volcanoes (Redoubt, Spurr, Iliamna, and Augustine), Alaska During the Period from 1990 to 1994."* U.S. Geological Survey Open-File Report 95-55
3. Symonds R.B., Mizutani Y. and Briggs P.H. (1996) *J. Volcanol. Getherm. Res., 73, 177.*
4. Symonds, R. B., Rose, W. I., Bluth, G. J. S., and Gerlach, T. M., 1994, In Volitiles in magmas, Caroll & Halloway, Eds., Mineralogical Society of America.
5. Toutain, J. P., Baubron, J.C., 1999, *"Gas geochemistry and seismo-tectonics: a review"*, Technophysics 304, 1-27
6. Watson I.M. et al., (2000), *"The relationship between degassing and ground deformation at Soufriere Hills volcano, Montserrat".,* J. Volcanol. Geoth.Res., 98, 117.
7. Zapata J.A., et al., (1997*) "SO2 fluxes from Galeras volcano, Columbia,1989-1995 : progressive degassing and conduit obstruction of a decade volcano."*, J. Volcanol. Getherm. Res., 77, 195.

CHAPTER 6
PROBABILITY CONTRIBUTION
FROM CRATER LAKES

Introduction

A volcanic lake is a cap of meteoric water over the vent of an active volcano. Only 12% of the world's 547 Holocene-aged (~10,000 yrs old) or younger volcanoes listed in the Catalog of Active Volcanoes of the World have such a lake. What makes volcanic lakes so rare is that their presence and expression require a special balance of volcanic heat flux versus atmospheric cooling and precipitation versus evaporation. The volcanic structure (e.g. substratum permeability and crater shape) is also important, but it is the former parameters- the endogene and exogene forces at work in a given crater subenvironment- that condition the lake's chemistry which in turn is indicative of volcanic activity and geo-hydrochemical processes. *Eruption Pro 10.6* will attempt to account for some of these contributions due to crater lake temperature monitoring.

Volcanic lakes are generally formed by one of three mechanisms: 1. explosive excavation (crater lakes) 2. collapse (caldera lakes) 3. blockage of common waterways (rivers, streams) by mudflows, lava flows or ash.

Volcanic lakes range in composition from common meteoric waters to those with a strong volcanic imprint, and temperatures can vary from ambient to close to boiling. Some subglacial lakes can be warmer than ice and have no direct contact with the atmosphere.

Lakes can only stably exist when the water influx = water outflux. This steady state condition can be determined by simple lake hydrology. In warm lakes, evaporation may play a significant role in determining the size or level of the lake at steady state. In such cases, one can gain insight into lake dynamics through energy budget analyses. For example, by calculating energy losses and gains from evaporation and radiation, it is possible to estimate the volcanic energy input into a lake.

Hazards Associated With Volcanic Lakes

Lahar Flows

Volcanic eruptions through lakes can cause failure of a retaining wall and a resulting flood of the lake waters. Many eruptions through lakes lead to massive lahar flows (hot mudflows), as is common at Kelut volcano in Indonesia. A tunnel was dug to that lake to control the lake level and volume to prevent major disasters.

Lake Explosions

True lake explosions may be thermal in origin (sudden catastrophic boiling event) or may be related to catastrophic degassing of the lake waters. Catastrophic degassing occurs in a special class of lakes (e.g., Monoun, Nyos, both in Cameroon) where the bottom waters have high dissolved CO_2 contents. A variety of triggers can cause an overturn of the lake, and a sudden decrease in hydraulic head may lead to explosive degassing of the lake waters. Almost 2000 people perished in 1986 during the Lake Nyos explosion. Most victims were asphyxiated in the cold CO_2 cloud that traveled as a gaseous density current from the lake down the valleys.

Jokuhlaups

In glaciated regions, subglacial eruptions can create large lakes that may escape through flood-channels causing glacier bursts or jokuh-laups, as happens periodically in Iceland.

Natural Pollution

Some very hot and concentrated crater lakes emit whitish clouds of HCl gas, forming acid aerosols in humid air. Many volcanic lakes leak acid and/or toxic fluids into the local watersheds where the water may be used for household or irrigation purposes. The environmental chemistry of volcanic lakes (natural pollution) is thus a rapidly emerging field. Fluorosis is a common disease in many areas with an abundance of volcanic fluids, and at the moment we can only guess at the effects of heightened levels of toxic elements such as the heavy metals, Li, As and Tl in local surface waters.

Algorithms

For the purposes of ***Eruption Pro 10.6***, the following guidelines are used as inputs:

If the crater lake temperature is reported as having a positive change of :

$<10^\circ$ C	= 0 probability contribution
$\geq \Delta 10^\circ - 19^\circ$ C	= 0.05 probability contribution
$\geq \Delta 20^\circ - 90^\circ$ C	= 0.10 probability contribution
$\geq \Delta 100^\circ$ C	= 0.20 probability contribution

References

1. R. Decker & B. Decker, 1981, "*Volcanoes*", Chapter 15

CHAPTER 7
PROBABILITY CONTRIBUTION DUE TO
THERMAL ANALYSIS

Introduction

Among some of the latest additions to the probability contribution suite is the input due to an increase on thermal output from the volcano under analysis. This accomplished through satellite based thermal imaging. Input for this contribution is obtained from the Geostationary Operational Environmental Satellite (GOES) namely, GOES-8, GOES-9, GOES-10, and GOES-12, the Operational Significant Events Imagery (OSEI), and the Advanced Very High Resolution Radiometer (AVHRR) satellite imaging. Images are analyzed, along with reports concerning the volcano being examined, and assigned a sliding scale probability contribution factor ranging from 0 to .1 probability. *Eruption Pro 10.6* will account for some of these contributions due to thermal imaging.

Imaging Types

GOES

GOES satellites provide the kind of continuous monitoring necessary for intensive data analysis. They circle the Earth in a geosynchronous orbit, which means they orbit the equatorial plane of the Earth at a speed matching the Earth's rotation. This allows them to hover continuously over one position on the surface. The geosynchronous plane is about 35,800 km (22,300 miles) above the Earth, high enough to allow the satellites a full-disc view of the Earth.

Because they stay above a fixed spot on the surface, they provide a constant vigil for the atmospheric "triggers" for severe weather conditions such as tornadoes, flash floods, hail storms, and hurricanes. When these conditions develop the GOES satellites are able to monitor storm development and track their movements.

GOES satellite imagery is also used to estimate rainfall during the thunderstorms and hurricanes for flash flood warnings, as well as estimates snowfall accumulations and overall extent of snow cover. Such data help meteorologists issue winter storm warnings and spring snow melt advisories. Satellite sensors also detect ice fields and map the movements of sea and lake ice. And they are used to detect heat from volcanoes about the globe. Figure 7-1 below illustrates a thermal detection of the crater Pu'u O'o on Hawaii's Kilauea volcano.

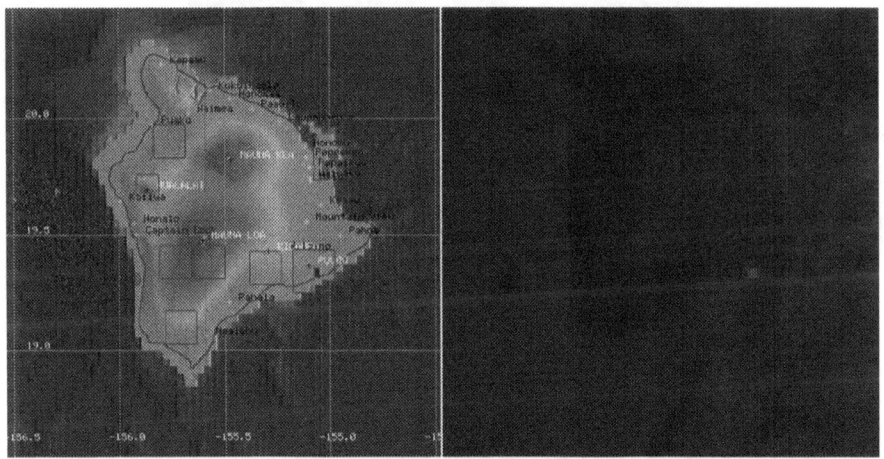

Figure 7-1. Example of a GOES Image of Hawaii (Pu'u O'o red thermal)

OSEI

The Operational Significant Event Imagery (OSEI)team produces high-resolution, detailed imagery of significant environmental events which are visible in remotely-sensed data available at the NOAA Science Center in Suitland, Maryland.

OSEI volcano products include multi-channel color composite imagery showing ash clouds (VSH), hotspots from lava flow (VIR) or both (VOL). We also create grayscale "split-window" images which use a channel differencing technique that enhances the appearance of ash clouds in imagery (DIF). This technique is often useful for distinguishing ash clouds from water vapour clouds.

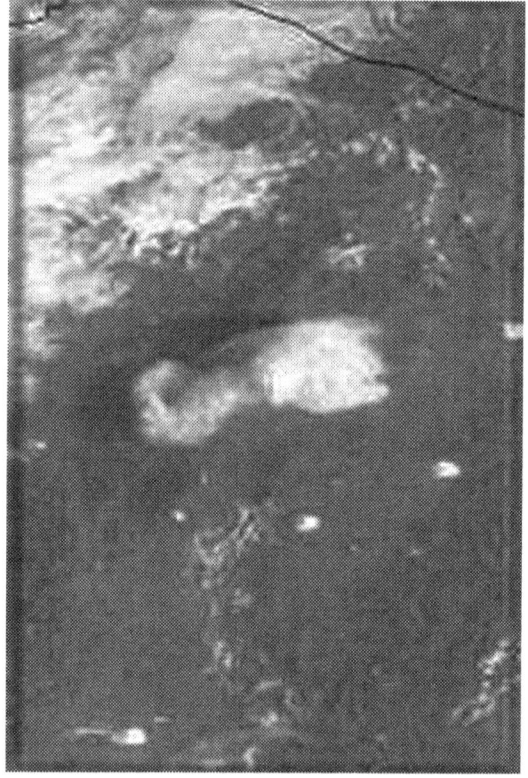

Figure 7-2. Example of OSEI imaging.

AVHRR

The Advanced Very High Resolution Radiometer (AVHRR) five channel scanning radiometer with 1.1-km resolution is sensitive in the visible and near-infrared, and the infrared 'window' regions. This instrument will be carried through NOAA-J (14); NOAA-K, L and M (15, 16, and 17) and will have a similar instrument with six channels and other improvements. AVHRR data are broadcast for reception by ground stations and also tape-recorded onboard the spacecraft for readout at the Fairbanks and Wallops Command Data Acquisition stations. These data can be recorded in 1.1-km resolution (the basic resolution of the AVHRR instrument) or at 4 km resolution; the swath width is >2600 km. The stored high resolution (1.1-km) imagery is known as Local Area Coverage (LAC). Owing to the large number of data bits, only about 11+ minutes of LAC can be accommodated on a single recorder. In contrast, 115 minutes of the lower resolution (4-km) imagery, called Global Area Coverage (GAC), can be stored on a recorder, enough to cover an entire 102 minute orbit of data.

The AVHRR has flown on the following U.S. civilian meteorological satellites: TIROS-N; NOAA-6 through NOAA-14, inclusive.

NOAA-K, L and M (NOAA-15 onwards) carry an enhanced version of the AVHRR scanner. It has six channels (three visible and three infra-red) but, for compatibility at receiving stations, only five are transmitted. Channel 3 is the visible channel during the daytime and the infra-red channel at nighttime. Additionally, the visible channels have been modified with a dual slope for calibration to give greater sensitivity at low light levels.

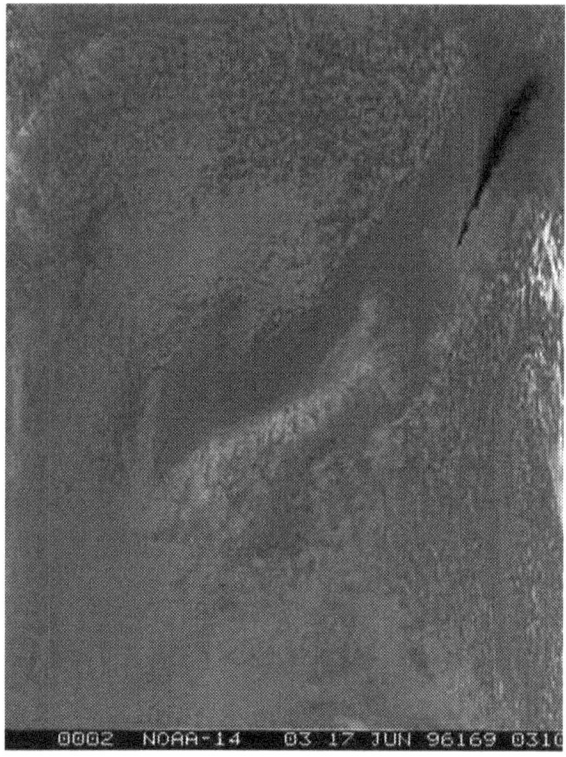

Figure 7-3. AVHRR Image of volcano Ruapehu, New Zealand.

Algorithms

The sliding scale probability is predominantly based on the colour interpretation of the examined images. e.g., a "light yellow" image would receive an assignment of .05 whereas a "red" image would receive a .1 probability value. Care is taken so as not to confuse reflectivity of clouds, etc. and otherwise "false" images from being interpreted as a thermal probability contribution.

No Thermal Colour	=	Probability of 0
Yellow Thermal Colour	=	Probability of .05
Red Thermal Colour	=	Probability of .10

References

1. AVHRR—Advanced Very High Resolution Radiometer, URL: http://www.sat.dundee.ac.uk/avhrr.html
2. NOAA/NCDC, URLhttp://www.ncdc.noaa.gov/oa/satellite/ satelliteseye/volcanoes/ruapehu96/ruapehu96.html
3. OSEI—Operational Significant Event URL: Imagery.RL:http://www.osei.noaa.gov/Events/Volcano/

CHAPTER 8
PROBABILITY CONTRIBUTION DUE TO
FREQUENCY OF ERUPTION ANALYSIS

Volcano Frequency of Eruption Analysis (VFEA)

Another of the important inputs to forecasting the probability of an eruption event is a look at the volcano's eruption frequency, i.e., what is the eruption rate or cycle if any? In the case of *Eruption Pro 10.6's* software package inputs, only the last ten years of a particular volcano is examined. For example, if a particular volcano (e.g., Kilauea in Hawaii) has erupted every year for the past 10 consecutive years, there is a high probability that it will erupt in the eleventh year. Volcanoes that have erupted at some time in the last 10 years, but not every year, will have their probability contribution lessened depending on how many times it has erupted within the 10 year period.

Eruption History & Volcano Explosivity Index (VEI)
History

The value of the recent volcano eruption record is obvious to volcanologists. The events of the present provide some of our best clues in interpreting the volcanic products of the past few billion years of the earth's history. Eye-witness accounts, photographs, and instrumental documentation all build a picture of the processes, rates, and interrelationships of events that cannot be found from products alone and yet are essential for understanding volcanism.

In studying a particular eruption, the accounts of similar historical eruptions can provide valuable guides in investigating a specific type of volcanic process or product. The historical record gives a useful dynamic context and in assessing the likelihood of future activity at a particular volcano. And lastly, of course, the historical behaviour of similar, but better instrumentally documented volcanoes, is often helpful. The eruption history of monitored volcanoes is loaded into the *Eruption Pro 10.6* database along with all the various input used for forecasting purposes.

Volcano Explosivity Index (VEI)

The Volcanic Explosivity Index (VEI) was devised by Chris Newhall of the U.S. Geological Survey and Steve Self at the University of Hawaii in 1982 to provide a relative measure of the explosiveness of volcanic eruptions.

Volume of products, eruption cloud height, and qualitative observations (using terms ranging from "gentle" to "mega-colossal") are used to determine the explosivity value. The scale is open-ended and ranges from 0, for non-explosive eruptions (less than 104 cubic metres of tephra ejected), to 8, for mega-colossal explosive eruptions that can eject 10^{12} cubic metres of tephra and have a cloud column height of over 25 km. Each interval on the scale represents a ten-fold increase in observed eruption criteria. Values higher than 8 can be determined if needed.

Note that ash, volcanic bombs, and ignimbrite are all treated alike—this is due to taking into account of the vesicularity (gas bubbling) of the volcanic products in question and the DRE (Dense Rock Equivalent) is calculated to give the actual amount of magma erupted. But one weakness is that the VEI does not take into account the magnitude of power output of an eruption. However, this is extremely difficult to detect with prehistoric or unobserved eruptions.

As an eruption of a volcano occurs, ***Eruption Pro 10.6*** calculates the new VEI average based on the loaded historical VEI average and the VEI determined for the new eruption. The default VEI when none is given within the eruption data from an authorized observatory or geo-agency is 2.

Table 8-1 below illustrates the classifications assigned to the VEI scale along with the volume of materials and height of plumes.

TABLE 8-1. VEI Classifications

VEI	PLUME HEIGHT	VOLUME	CLASSIFICATION	EXAMPLE
0	<100 m	1000s m^3	Hawaiian	Kilauea
1	100-1000 m	10,000s m^3	Hawaiian/Strombolian	Stromboli
2	1-5 km	1,000,000s m^3	Strombolian/Vulcanian	Galeras ('92)
3	3-15 km	10,000,000 m^3	Vulcanian	Ruiz (1985)
4	10-25 km	100,000,000s m^3	Vulcanian/Plinian	Galunggung
5	>25 km	1 km^3	Plinian	St. Helens
6	>25 km	10s km^3	Plinian/Ultra-Plinian	Krakatau
7	>25 km	100s km^3	Ultra-Plinian	Tambora
8	>25 km	1000s km^3	Ultra-Plinian	Toba

Algorithms

The most improved probability contribution is due to the analysis of a volcano's frequency of eruptions. Since the conception of ***Eruption Pro*** (1989), the Volcano Frequency of Eruption Analysis (VFEA) has been tracked. This contribution uses the simplest of probability models namely;

$$P = \frac{k}{h}$$

where   k = # of outcomes (eruptions in the last 10 years)
          h = # of possible outcomes (last 10 years)

For example, since 1996, volcano Hekla, located in Iceland, has erupted once. Using the equation above, this means that the probability contribution for volcano Hekla is .100 for the current year's forecast.

$$P = \underline{k} = \underline{1} = \quad .100$$
$$h \quad 10$$

References

1. Trombley, R. B., 1995, *"A Computer Based Long-Range Volcano Eruption Forecasting Programme, "Eruption""*, EOS, Transactions, American Geophysical Union (AGU) Vol 76, No. 46, 7 Nov 1995 Fall Convention. Revised May, 1996.
2. Wikipedia, the Free Encyclopedia, URL: http://en.wikipedia.org/wiki/Volcanic_Explosivity_Index

CHAPTER 9
PROBABILITY CONTRIBUTION DUE TO LUNAR INFLUENCES

Introduction

While reviewing journals of scientific observations and at the notion of other scientists, e.g., Dr. Steve O'Meara, et al, there seems to be a noticeable correlation between increasing volcanic activity and lunar cycles. Other observers throughout history have noticed the possibility of such a connection, but always as a footnote and always when looking back at eruptions that have already occurred. To date, no one seems to have given the matter comprehensive study and no one has attempted to employ some of these lunar patterns as one of the inputs to forecast volcanic eruptions. ***Eruption Pro 10.6*** will attempt to account for some of these contributions due to lunar effects. One must remember that the moon has an obvious effect on the water of the earth due to tidal influences and magma is also a liquid!

Tidal Forces Associated With The Moon

In the following algorithms, the effect of the earth's rotation about its own axis will be ignored. The effect produced by this rotation can be taken into account in the expression for gravity. Thus we can regard the earth merely as moving without rotation along a circular path relative to the earth/moon system. Every point of the inter-rotational body depicted in Figure 9-1 is subject to the same centripetal acceleration since the orbits of each point are of identical radii. The acceleration is given by:

$$a_e = \frac{G\,M_m}{d^2}$$

Inserting the values of G, M_m, and d from Lunar Calculations section (Appendix B) we obtain

$$a_e = 3.32E+01$$

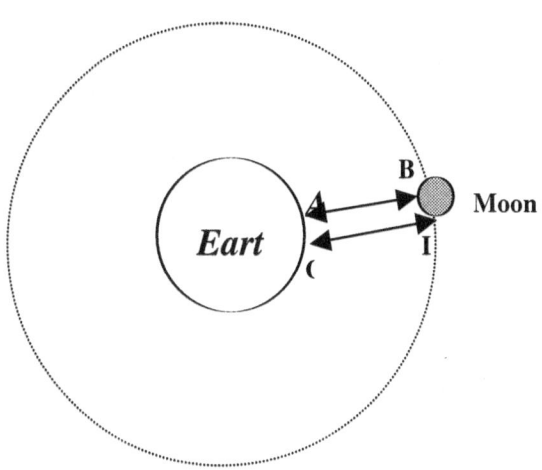

Figure 9-1. Inter-rotational body mechanics.

If we consider the concept of centrifugal force per unit mass to replace the centripetal acceleration, then at Point A on the earth's surface the force of attraction on a unit mass due to the moon is partially offset by centrifugal force on the unit mass at this point. The net force per unit mass at A is

$$f_A = G M_m \left[\frac{1}{(d - \alpha)^2} - \frac{1}{d^2} \right]$$

and this is directed towards the moon. The centrifugal force per unit mass at Point B is the same but the attractive force due to the moon is less. The net force per unit mass at B is

$$f_B = G M_m \left[\frac{1}{d^2} - \frac{1}{(d + \alpha)^2} \right]$$

and this is directed away from the moon. The expressions above can be written as

$$f_A \approx f_B \approx G M_m \frac{2\alpha}{d^3}$$

with no sensible loss of accuracy. Now, if we use the approximation

$$g \approx G M_e / \alpha^2$$

then

$$f_A \approx f_B \approx 2 \frac{M_m}{M_e} \left[\frac{\alpha}{d} \right]^3 g$$

Inserting the values of M_m, M_e, α and d from Lunar Calculations section (Appendix B) we obtain

$$f_A = f_B = 1.06E\text{-}24 \ g$$

Thus at the points a and b the net effect is to produce a reduction of gravity in the amount of 0.00011 cm/sec only.

The lunar tidal force at other locations can also be analyzed.

The distance from point P on the earth to the moon's centre is given by

$$r^2 = \alpha^2 + d^2 - 2\,\alpha\,d\,\text{Cos}\,\theta$$

where θ is the angle formed by the lines joining the centre of the earth with P and with the moon's centre. The components of the attractive force of the moon in the directions parallel and normal to the line joining the centres of the earth and moon are, respectively,

$$f_\varepsilon = G\,\underline{M}_m\,(1-\alpha^2)$$
$$\,r^2$$

$$f_\eta = G\,\underline{M}_m\,\alpha$$
$$\,r^2$$

where

$$\tan\alpha = \frac{\alpha\,\text{Sin}\,\theta}{(d-\alpha\,\text{Cos}\,\theta)}$$

Because of the very small value of the ratio α/d (about 1/60), the following approximations are in order:

$$f_s = G\,\underline{M}_m$$
$$\,r^2$$

$$f_\eta = G\,\underline{M}_m\,\frac{\alpha}{d}\,\text{Sin}\,\theta$$
$$\,r^2$$

The net value of the two forces acting parallel to the line OL is

$$f_{s'} = G\,M_m \left[\frac{1}{r^2} - \frac{1}{d^2} \right]$$

and

$$f_{s'} = G\,M_m\,\frac{2\alpha\,\mathrm{Cos}\,\theta}{r^2\,d} \left[1 - \frac{\alpha\,\mathrm{Cos}\,\theta}{2d} \right]$$

The net force components can be written approximately as

$$f_s = G\,M_m\,\frac{2\alpha}{d^3}\,\mathrm{Cos}\,\theta$$

$$f_\eta = G\,M_m\,\frac{\alpha}{d^3}\,\mathrm{Sin}\,\theta$$

It is more convenient to deal with the radial and tangential components and of the net tidal force at the earth's surface.

$$f_s = G\,M_m \left[\frac{2\alpha}{d^3}\,\mathrm{Cos}^2\,\theta - \frac{\alpha}{d^3}\,\mathrm{Sin}^2\,\theta \right]$$

$$f_\theta = G\,M_m\,\frac{3\alpha}{d^3}\,\mathrm{Cos}\,\theta\,\mathrm{Sin}\,\theta$$

Introducing the approximation and simplifying leads to

$$f_s = \frac{1}{2} G \frac{M_m}{M_e} \left[\frac{\alpha}{d} \right]^3 (1 + 3 \cos 2\theta)$$

$$f_s = \frac{3}{2} G \frac{M_m}{M_e} \left[\frac{\alpha}{d} \right]^3 \sin 2\theta$$

When $\theta = 0$, and reduces to the above. For $\theta = 90°$

$$f_s = -G \frac{M_m}{M_e} \left[\frac{\alpha}{d} \right]^3 = -.000055 \text{ dynes/gm// } f_\theta = 0$$

indicating a slight increase in gravity (negative being towards the earth's centre).

Actually, it is the tangential (horizontal) component which leads to the development of the tides since the slight effect on g by the vertical force has virtually no effect. The maximum tide-producing force associated with the moon occurs at a value of and has the value

$$(f_\theta)_{max} = .0000833 \text{ dynes/gm}$$

This force is equivalent to the component of gravity acting along a slope of 8 cm per 1000 km which is about twice in magnitude the mean slope of the sea surface along the equator in the Pacific, but is of the order of one fourth the slope of the surface across the major currents such as the Gulf Stream and Kuroshio Current. Consequently, the horizontal tide force is far from negligible compared with other horizontal forces in the sea.

Use Of Hypergeometric Distribution

Hypergeometric probability is determined to be the best probability distribution to be used with lunar influences as the hypergeometric distribution is used to determine the probability of obtaining a certain sample outcome when sampling without replacement. The hypergeometric probability function is described as

$$f(x) = \frac{\binom{r}{x}\binom{N-r}{n-x}}{\binom{N}{n}}$$

where r = number of elements in the population labeled successes, x = # of successes, n = number of trials, N = number of elements in the population, and f(x) = the probability of x successes in n trials.

Hypergeometric Calculations & Algorithm

$$f(x) = P_{lunar} = \frac{\binom{r}{x}\binom{N-r}{n-x}}{\binom{N}{n}}$$

Using the above equation, the probability contribution to be used in *Eruption Pro 10.6* for the probability contribution due to lunar influence is as follows:

For x = 1, N = 29.53059, r = 1, n = 365.25

$$P_{lunar} = 0.079452055$$

References

1. Trombley, R. B., 1995, "*A Computer Based Long-Range Volcano Eruption Forecasting Programme, "Eruption"*", EOS, Transactions, American Geophysical Union (AGU) Vol 76, No. 46, 7 Nov 1995 Fall Convention. Revised May, 1996.
2. Wolfram Research, Astronomy, Lunar Cycles. URL: http://scienceworld.wolfram.com/astronomy/LunarCycles.html

CHAPTER 10
PROBABILITY CONTRIBUTION DUE
TO SOLAR INFLUENCES

Introduction

While reviewing journals of scientific observations and at the notion of other scientists, e.g., Dr. T. A. Jaggar of the Hawaiian Volcano Observatory (HVO), there seems to be a noticeable correlation between increasing volcanic activity and the sun. Other observers throughout history have noticed the possibility of such a connection, but always as a footnote and always when looking back at eruptions that have already occurred. To date, no one seems to have given the matter comprehensive study and no one has attempted to employ some of these solar patterns as one of the inputs to forecast volcanic eruptions. *Eruption Pro 10.6* will attempt to account for some of these contributions due to solar effects albeit they are small contributions.

Tidal Forces Associated With The Sun

The tidal forces on the earth with attraction of the sun can be derived in a manner entirely similar to that of the moon. In this case, then motion of the centre of the earth-moon system in its orbit about the sun must be taken into account. This leads to a centrifugal force per unit mass at any point on earth is equal to

$$a_e = \frac{G\,M_s}{D_s^{\,2}}$$

where M_s is the mass of the sun and D_s is the distance between the earth-moon centre and the sun. This distance varies by about 3 percent during the course of the year due to the eccentricity of the centre of the orbit from the centre of the sun. However, as a first approximation D_s may be regarded as a constant.

Using the above equation yields (See Solar Calculations in Appendix C).

$$a_e = 5.90E+03$$

The final expressions for the vertical and horizontal tidal forces associated with the sun are:

$$f_r = \frac{1}{2} \, G \, \frac{M_s}{M_e} \left[\frac{a_e}{d} \right]^3 (1 + 3 \, \text{Cos} \, 2\theta)$$

$$f_\theta = \frac{3}{2} \, G \, \frac{M_s}{M_e} \left[\frac{a_e}{D_s} \right]^3 \text{Sin} \, 2\theta$$

The greater distance from the earth to the sun than from the earth to the moon offsets the effect of the greater mass of the sun. By inserting the appropriate constants we find that

$$E_s = \frac{M_s}{M_e} \left[\frac{a_e}{D_s} \right]^3 = 2.02E-08$$

which is about 0.46 times that for the moon (See Solar Calculations in Appendix C). Thus, the tide producing effect of the sun is roughly half that due to the moon.

The angle θ' above represents the angle between the line joining the centre of the earth with a point on the earth's surface and a line between the centres of the earth and sun. For a given point on the earth the angles θ and θ' have different values depending upon the relative positions of the moon and the sun at a given time. Furthermore, at a particular point on the earth's surface the angles θ and θ' change periodically with time in a rather complicated manner. This is due to the combined effect of the rotation of the earth about an axis inclined to the earth/sun ecliptic plane and the varying aspects of the moon and sun to earth. In addition to these effects, the plane of the moon's orbit is inclined somewhat (about 5 degrees) to that of the earth/sun ecliptic plane, leading to an additional complicating feature with regard to the possible combinations of the angles θ and θ'.

Use Of Hypergeometric Distribution

Hypergeometric probability is determined to be the best probability distribution to be used with solar influences as the hypergeometric distribution is used to determine the probability of obtaining a certain sample outcome when sampling without replacement. The hypergeometric probability function is described as

$$f(x) = \frac{\binom{r}{x}\binom{N-r}{n-x}}{\binom{N}{n}}$$

where r = number of elements in the population labeled successes, x = # of successes, n = number of trials, N = number of elements in the population, and f(x) = the probability of x successes in n trials.

Hypergeometric Calculations & Algorithm

$$f(x) = P_{solar} = \frac{\binom{r}{x}\binom{N-r}{n-x}}{\binom{N}{n}}$$

Using the above equation, the probability contribution to be used in *Eruption Pro 10.6* for the probability contribution due to lunar influence is as follows:

For x = 1, N = 365.25, r = 1, n = 1

$$P_{solar} = 0.00274$$

Considering Sunspots

As was stated previously, while reviewing *"The Volcano Letter"*, No 326, 26 March 1931, in a study completed by Dr. T. A. Jaggar, he points out that there seems to be a definite correlation between curves of frequency of sunspots and curves of frequency of volcanic eruptions. There has also been contradictory results published as well. Jaggar found that sunspot cycle is 11.1 years in length and that was the basis for the analysis.

Why Sunspots?

If the reader asks why sunspots should have anything to do with volcanoes, the answer is that nobody really knows. Times of maximum sunspots affect radio reception on the earth, magnetism on the earth, and auroras in the arctic regions. Sunspots are accompanied by gigantic eruptions of gas on the sun and colossal electrical phenomena in the solar system. If earth magnetism and electricity are in some way associated with gravity, volcanism may be affected. If heat from the earth's radio-activity affects volcanism, the sun may in turn affect the earth's radiations. Finally, if volcanic emanations on the earth are a last remnant of solar processes here, those processes by unknown means may be sympathetic with the sun. **Eruption Pro 10.6** will attempt to account for some of these contributions due to sunspot effects.

Sunspot Mathematics:

Using 1/11.1 as a basis, then

$$\text{Per year} = (1/11.1)/10 = 0.009009$$

and factoring in the years of sunspot maximum and minimums, the following table will be used as a guide for input to **Eruption Pro 10.6** as input for probability contribution due to sunspots.

TABLE 10-1. Sunspot Cycle and Contribution Factors*

Per year = 0.009009
(1/11.1 = 0.0909009)

Year	Factor	Year	Factor
2006	0.00901	2017	0.00901
2007	0.00271	2018	0.00271
2008	0.01802	2019	0.01802
2009	0.02703	2020	0.02703
2010	0.03604	2021	0.03604
2011	0.04505	2022	0.04505
2012	0.05405	2023	0.05405
2013	0.06306	2024	0.06306
2014	0.07207	2025	0.07207
2015	0.08108	2026	0.08108
2016	0.09009	2027	0.09009

* = Probability contribution

Hypergeometric Distribution Function Calculations & Algorithm For Sunspots

It is also possible to use the following calculation in lieu of the above table.

$$f(x) = P_{sunspot} = \frac{\begin{pmatrix} r \\ x \end{pmatrix} \begin{pmatrix} N-r \\ n-x \end{pmatrix}}{\begin{pmatrix} N \\ n \end{pmatrix}}$$

For x = 1, N = 11.1, r = 1, n = 4054.275

$$f(x) = P_{sunspot} = 0.00271$$

References

1. National Geophysical Data Center)NGDC), URL: http://www.ngdc. noaa.gov/stp/SOLAR/SSN/ssn.html

2. Trombley, R. B., 1995, "*A Computer Based Long-Range Volcano Eruption Forecasting Programme, "Eruption""*, EOS, Transactions, American Geophysical Union (AGU) Vol 76, No. 46, 7 Nov 1995 Fall Convention. Revised May, 1996.

CHAPTER 11
REPOSE ANALYSIS

Introduction

Repose is defined as the interval of time between eruptions. Although a repose analysis is not one of the probability contributions to **Eruption Pro 10.6**, it nevertheless can be a useful analysis. This is particularly true when an "educated guess" as to when the next eruptive event of a volcano may occur. It is included as one of the outputs of the software package **Eruption Pro 10.6**. It is also interesting to compare the repose analysis projected years until an eruption event to the forecast(s) as produced by **Eruption Pro 10.6**.

Methodology

As part of the results displayed, **Eruption Pro 10.6** performs a repose analysis of every volcano within its database. From the loaded historical eruption data for each volcano, the first goal of the repose analysis is to determine the mean of the interval of time between eruptions. Next, the number of reposes as read from the database for a particular volcano must be determined to determine the use of small (<30 reposes) or large (≥30 reposes) sample statistics. Then the standard deviation is determined, again from the volcano database. Based on whether there are small or large sample sizes of repose numbers, the sampling error is determined. Then the probability interval is calculated and finally, the interval calculated is subtracted from the mean for the low interval and added to the mean for the high interval. This results in the determination of the projected years maximum until an eruptive event occurs and the forecast year limit.

Mathematics

The statistical mean of the number of reposes is determined by

$$X_{bar} = \Sigma X_i / n$$

where X_i = Repose #, n = # of reposes, X_{bar} = mean of repose

The standard deviation of the reposes is determined by

$$\sigma_{xbar} = [\ \Sigma(X_i - X_{bar})^2 / (n - 1)\ ]^{.5}$$

Where X_i = Repose #, n = # of reposes, X_{bar} = mean of repose, σ_{xbar} = standard deviation.

We use for large sample sizes a 95% confidence interval figure of 1.96 ($z_{\alpha/2}$). This means that 95% of the values of a normally distributed random variable lie within ±1.96 standard deviations from the mean. Therefore the probability sampling error is expressed as

$$z_{\alpha/2}\ \sigma_{xbar}$$

and the probability interval is formed by

$$X_{bar} \pm z_{\alpha/2}\ \sigma_{xbar}$$

With respect to small sample statistics where the number of reposes is less than 30, the calculation of the mean and standard deviation are the same as for large sample sizes. However, the probability sampling error is different. For small sample size, the probability sampling error expressed as

$$t_{\alpha/2}\ \sigma_{xbar}$$

where $t_{\alpha/2}$ is representing the t distribution which is actually a family of probability distributions similar to $z_{\alpha/2}$ distribution with a specific t distribution depending on a parameter known as *degrees of freedom* (DOF). **Eruption Pro 10.6** has a built in DOF look-up table fro reposes less than 30. DOF is calculated simply as follows:

$$DOF = NOR - 1$$

where NOR = number of reposes.

Algorithms

Software package **Eruption Pro 10.6** reads from the database, the number of reposes, repose mean, and standard deviation, last eruption year, and the current year. The following examples are then calculated from the read data:

Large Sample Size Example: **SLAMET (Java)**

NOR = 44, σ_{xbar} = 8.846, CI = 1.96, X_{bar} = 5.305 and LEY = 2000

RINT = CI * (σ_{xbar} / NOR$^{\frac{1}{2}}$) = 1.96 * (8.846 / 6.633)		=	**2.614**
RIL = X_{bar} - RINT = 5.305 – 2.614		=	**2.691**
RIH = X_{bar} + RINT = 5.305 + 2.614		=	**7.919**
RFYR = LEY + RIH = 2000+ 7.919		≈	**2008**
RMYRS = RFYR - CYEAR = 2008 - 2006		=	**2**

where RINT = The Repose interval, CI = Confidence Interval, σ_{xbar} = Standard Deviation, RIL = Lower interval value, RIH = Upper interval value, LEY = Last Eruption Year, RFYR = Forecasted year limit, RNYRS = Years until limit is reached, and CYEAR = Current Year.

Small Sample Size Example: **BAYONNAISE (Japan)**

NOR = 20, σ_{xbar} = 5.667, TI = 2.093, X_{bar} = 5.50 and LEY = 1988

DOF = NOR – 1	**= 20 – 1**	**=**	**19**
Therefore (from DOF Table)	**TI**	**=**	**2.093**
RINT = CI * (σ_{xbar} / NOR$^{1/2}$) = 2.093 * (5.667 / 4.472)		**=**	**2.652**
RIL = X_{bar} - RINT = 5.500 – 2.652		**=**	**2.848**
RIH = X_{bar} + RINT = 5.500 + 2.652		**=**	**8.152**
RFYR = LEY + RIH = 1988+ 8.152		**≈**	**1996**
RMYRS = RFYR - CYEAR = 1996 - 2006		**=**	**-10**

where RINT = The Repose interval, TI = Confidence Interval, σ_{xbar} =
    Standard Deviation, RIL = Lower interval value, RIH = Upper
    interval value, LEY = Last Eruption Year, RFYR = Forecasted
    year limit, RNYRS = Years until limit is reached, and CYEAR =
    Current Year.

One will notice that in this case (Bayonnaise), the years until the limit is
reached, is a negative number. ***Eruption Pro 10.6*** then presents a special alert, in one of the available options. The special alert is an
"Overdue" alert to alert to the fact that the volcano selected (in this case
Bayonnaise) is 10 years overdue for an eruption.

References

1. Anderson, et al, 5th Ed., 1993, *Statistics for Business and Economics*, Chap. 8, pgs.252ff
2. Trombley, R. B., 2000, T. A. Jackson (Ed), *Caribbean Geology— Into The Third Millenium,* Chapter 23.

CHAPTER 12
RESULTS, RELIABILITY & SHORTCOMINGS

Results

The original *"Eruption"* had limited, yet somewhat successful beginnings back in 1989 when it was first released. Further research and suggestions from colleagues about the field, e.g., Drs. Jonathan Dehn of University of Alaska, Jean-Paul Toutain, Université Paul Sabatier, et al, have lead to improved success in the ability of this software to forecast long-range volcanic eruptions of all type of volcanoes except submarine type volcanoes. Since the original software development, several new and improved algorithms and corrections have been incorporated into *"Eruption"* with much better results than from the years 1989 through 1995. Not only is the historical data taken into account but also the real-time seismic, deformation and COSPEC analysis from reporting volcanoes. *"Eruption"* was able to combine these factors into making long-range forecasts taking into consideration all of these factors. Progress continued to be slow, but forward. In 1997, new algorithms have been incorporated into the software and resulted in some improvements. Finally, in August of 2005, the latest new and improved algorithms were incorporated into what is now *Eruption Pro 10.6*. These changes included the accounting of the effects, albeit very small, of both lunar and solar, including sunspots, activity. The present version of *Eruption Pro 10.6* now takes the following into account: Historical and current available eruption data, near real-time measurement data including, Seismic, Deformation, Thermal, Frequency of Eruptions, Solar & Lunar influences, Crater Lake Temperature (if applicable), and COSPEC.

All versions of *"Eruption"* generate three types of forecasts; a statistically projected next eruption year, the next forecasted beginning eruption year with an ≥50% probability of eruption occurrence, and finally, the next forecasted beginning eruption year with an ≥95% probability of eruption occurrence.

Table 12-1 below illustrates the results (to date) of all the versions of *"Eruption"*. One will note that after 1997, the results dramatically improved due to the incorporation of the above noted improvements.

TABLE 12-1. Eruption Pro Annual Results

	FORECASTS			≥50% FORECASTS			≥95% FORECASTS			# OF	Eruption's
YEAR	FE	AE	PCT	5FE	5AE	PCT	9FE	9AE	PCT	EVENTS	Accuracy
1989	2	0	0.0	18	17	94.4	5	4	80.0	40	52.50
1990	3	2	66.7	5	4	80	0	0	100.0	26	23.08
1991	3	2	66.7	12	10	83.3	5	5	100.0	27	62.96
1992	6	0	0.0	2	1	50	5	4	80.0	39	12.82
1993	6	2	33.3	8	8	100	3	1	33.3	37	29.73
1994	5	0	0.0	13	11	84.6	1	0	0.0	39	28.21
1995	0	0	100.0	4	3	75	1	1	100.0	38	10.53
1996	11	7	63.6	15	7	46.7	8	5	62.5	31	61.29
1997	18	11	61.1	14	7	50	8	6	75.0	28	85.71
1998	23	12	52.2	26	15	57.7	6	5	83.3	38	94.12
1999	16	13	81.3	34	26	76.5	7	5	71.4	50	93.62
2000	21	10	47.6	37	31	83.8	8	6	75.0	56	90.39
2001	12	6	50.0	39	29	74.4	9	5	55.6	49	90.91
2002	15	9	60.0	45	34	75.6	6	3	50.0	55	92.00
2003	13	6	46.2	43	30	69.8	4	2	50.0	47	90.70
2004	14	9	64.3	49	38	77.6	6	4	66.7	58	91.07
2005	15	10	66.7	56	43	76.8	5	5	100.0	63	95.08
2006	23	16	69.6	43	30	69.7	4	4	100.0	55	94.34

Main *Eruption Pro 10.6* Output Example

There are many outputs that *Eruption Pro 10.6* is capable of performing. The following examples below are typical outputs of the software. The first example presented is an example out put of a volcano that has not as yet erupted in 2006, Oshima in Japan. The second example is of a volcano that has erupted, in this case 2006, Pacaya in Guatemala. Some of the terms, abbreviations, etc. may be looked up in Appendix D.

In the Repose Analysis section, the analysis is done starting with the reposes since the year 0 A.D. This is done due to the fact that prior to this year, the accurate record of actual eruption dates is sometimes subject to interpretation such as those determined by dendrochronology (tree ring analysis). Since year 0 A.D., very accurate records can be afforded and are therefore incorporated into the software package algorithms.

TABLE 12-2. *Eruption Pro 10.6* Output For Volcano Oshima

Volcano:	OSHIMA	STRATO	
Location:	JAPAN	34.72N	139.40E
Elevation:		758 m	2,487 ft
Number of Eruptions:	82	Avg VEI = 2	
Year of 1st Eruption:	-80		
Year of Last Eruption:	1990		
U =	3.988327E-02		
Total Years:	2056	VEI Determined For This	
Pr(0) =	.8434461	Eruption = Unknown	
Pr(1) =	.1565539		
PSI =	0		
SOLAR =	.00274		
SSPOT	.03604		
LUNAR =	.07945		
PTERM =	0		
PDEF =	0	REPOSE ANLAYSIS	
PSO2 =	0.000	(Since year 0 A.D. @95% C/L)	
VFEA Code =	0	Avg. Repose =	23.457 yrs
Thermal Imaging Available:	NO	C/L Interval =	8.226 yrs
Microgravity Monitored:	NO	Forecasted Year Limit =	2022
Statistical Forecasted Eruption Year =	2063		
Year of Forecasted Eruption @ ≥50% =	2007	Projected Yrs. Max.	
Year of Forecasted Eruption @ ≥95% =	2065	Until Eruption:	16

TABLE 12-3. *Eruption Pro 10.6* Output For Volcano Pacaya

Volcano:	PACAYA	COMPLEX
Location:	GUATEMALA	14.38N 090.60W
Elevation:		2,552 m 8,373 ft
Number of Eruptions:	36	Avg VEI = 1.963
Year of 1st Eruption:	750	
Year of Last Eruption:	2006	
U =	2.866242E-02	
Total Years:	1256	VEI Determined For This
Pr(0) =	0	Eruption = 2.000
Pr(1) =	1	
PSI =	0	
SOLAR =	.00274	
SSPOT	.03604	
LUNAR =	.07945	
PTERM =	0	
PDEF =	0	REPOSE ANLAYSIS
PSO2 =	0.000	(Since year 0 A.D. @95% C/L)
VFEA Code =	1	Avg. Repose = 32.205 yrs
Thermal Imaging Available:	YES	C/L Interval = 0.000 yrs
Microgravity Monitored:	YES	Forecasted Year Limit = 0
Statistical Forecasted Eruption Year =	2006	
Year of Forecasted Eruption @ ≥50% =	2032	Projected Yrs. Max.
Year of Forecasted Eruption @ ≥95% =	2110	Until Eruption: 0

Reliability Analysis

As with any new or experimental software package, particularly in the case of volcanic eruption forecasting, there is a need to perform an analysis to determine if the model is accurate and if the methodology proposed has sound footing. A reliability analysis was performed per se in the original paper on *Eruption Pro* with mixed accuracy results (See Table 12-4 for accuracy results.) However, a retroactive model testing analysis has been performed for 1998 thru 2006 which reveals (to date) a 94.34% overall accuracy level.

Formally, reliability is defined as the *probability that a system performs its intended function for a stated period of time under specified operating conditions.* This definition has four important elements: probability, time, performance and operating condition. First, reliability, with respect to its application to **Eruption Pro 10.6**, is a probability value between 0 and 1. Thus, it is a numerical measure that has a precise meaning. Expressing reliability in this way provides for a valid basis for comparison of different **Eruption Pro 10.6** designs. The second element is time. The time factor governing **Eruption Pro 10.6's** is the period of one year. i.e., **Eruption Pro 10.6's** ability within a period of a calendar year to forecast. Thirdly, is the element of performance. This refers to the objective for which **Eruption Pro 10.6** was designed. The term *failure* is used when the performance of **Eruption Pro 10.6** is not met. And, in this case, a reliability failure is noted if **Eruption Pro 10.6** does not perform within a given year. Fourth, and lastly, the operating conditions under which **Eruption Pro 10.6** is operating is the analysis of historical as well as the near real-time data for each volcano that **Eruption Pro 10.6** performs an analysis.

In terms of the reliability analysis for the programme **Eruption Pro 10.6**, let us first consider the failure rate. Failure rate, as applied to this software programme, is determined by the number of failures (times **Eruption Pro 10.6** fails to properly forecast an erupting volcano in the given year) per unit of time (1 year). This may be represented by:

$$\lambda = \frac{NOF}{(NOV) * (NOY)}$$

where

NOF = Number of failures
NOV = Number of volcanoes forecasted
NOY = Number of years

If λ is the failure rate, the probability density function representing failures is given by the exponential density

$$f(t) = \lambda\, e^{-\lambda T}, t \geq 0$$

Specifically, the probability of failure in the interval (0, T) is given by the cumulative distribution function:

$$F(T) = 1 - e^{-\lambda T}$$

Since the reliability is the probability of *correct forecasting* by *Eruption Pro 10.6*, then the reliability function is given by:

$$R(T) = 1 - F(T) = e^{-\lambda T}$$

This represents the probability that *Eruption Pro 10.6* will correctly forecast within **T** units of time. **T** units of time, in the case of *Eruption Pro 10.6's* operating environment, is one year.

Table 12-4 illustrates the calculated reliability analysis for the years 1998 through 2006 (to date at this writing). In the previous versions of *Eruption Pro*, the analysis yielded an average, considering all years from 1989 through 1997, an overall reliability of 71.7%. This means that previous versions of *Eruption Pro's* reliability to correctly & accurately forecast volcanic eruptions will be incorrect 28.3% of the time (or would be correct 71.7% of the time). Today's version of *Eruption Pro* has a reliability of 93.8%. This means that current version of *Eruption Pro's* reliability to correctly & accurately forecast volcanic eruptions will be incorrect only 6.2% of the time (or would be correct 93.8% of the time).

Again, since the incorporation of improved algorithms, thusfar, the 2006 reliability analysis of *Eruption Pro 10.6's* forecasting ability indicates that the all combined and improved algorithms (put forth in the software in late 1996, 1998 and again in 2005) are in fact working well.

TABLE 12-4. Reliability Analysis 1998-2006

Year	# of Years	# of Volcanoes	# of Failures	λ	f(t)	F(T)	R(T)
1998	1	38	3	0.079	0.073	0.076	0.924
1999	2	51	3	0.059	0.055	0.057	0.943
2000	3	56	4	0.071	0.067	0.069	0.931
2001	4	49	4	0.082	0.075	0.078	0.922
2002	5	55	4	0.073	0.068	0.070	0.930
2003	6	47	1	0.021	0.021	0.021	0.979
2004	7	58	5	0.086	0.079	0.083	0.917
2005	8	63	3	0.048	0.045	0.047	0.953
2006	9	55	3	0.055	0.052	0.053	0.947
Totals:		472	30	0.064	0.060	0.062	0.938

Shortcomings

Even with the success of ***Eruption Pro 10.6***, there are a few shortcomings of the software. Arguably, the largest is the lack of available data on all the volcanoes that the software monitors. It is however, an unrealistic expectation to have all volcanoes, (particularly those volcanoes that are very remote), monitored to the degree necessary for very accurate forecasting results. There is simply not the equipment, money and personnel to encompass such an endeavour. Another shortcoming is the timeliness of current acquired data. In some cases, the data received and confirmed on various volcanoes is almost immediate, in other cases, it may be a day or two before data is received and confirmed.

There are other probability contributions that may be considered as potential inputs to ***Eruption Pro 10.6*** but are not currently in the software. For example, the monitoring and analysis of other types of output gasses is another area of probability contribution that may be considered although SO_2 analysis is currently monitored and input. Lastly, the analysis of volcanic plumes, and the measurement of gas fluxes in plumes, may also be a viable contribution. Much work is being done in this area, particularly by colleague, Dr. Jean-Paul Toutain.

References

1. Evans, J. R., Lindsay, W. M., 1993, *"The Management and Control of Quality"*, Chapter 15, pgs. 502ff
2. Trombley, R. B., 1995, *"A Computer Based Long-Range Volcano Eruption Forecasting Programme, "Eruption""*, EOS, Transactions, American Geophysical Union (AGU) Vol 76, No. 46, 7 Nov 1995 Fall Convention. Revised May, 1996.
3. Trombley, R. B., 1990, *"Computer modeling of statistical explosive patterns and the probability of volcanic event forecasting."*, Digital Equipment Corporation, U.S. Education Services, white paper.
4. Trombley, R. B., 2000, T. A. Jackson (Ed), *Caribbean Geology— Into The Third Millenium,* Chapter 23.

CHAPTER 13
FORECASTING OF SUPERVOLCANOES

Introduction

The term "supervolcano" has no specifically defined scientific meaning. It was used by the producers of a British TV programme in 2000 to refer to volcanoes that have generated Earth's largest volcanic eruptions. As such, a supervolcano would be one that has produced an exceedingly large, catastrophic explosive eruption (VEI = 8) and a giant caldera. Because Yellowstone has produced three such very large caldera-forming explosive eruptions in the past 2.1 million years, the producers considered it to be a supervolcano.

Because there is no well-defined minimum size for a "supervolcano," there is no exact number of such volcanoes. Examples of volcanoes that produced exceedingly voluminous pyroclastic eruptions and formed large calderas in the past 2 million years would include Yellowstone, Long Valley in eastern California, Toba in Indonesia, and Taupo in New Zealand. The forecasting of eruptions of this type of volcano presents extremely difficult and interesting new problems for the volcanologist to solve.

Some 640,000 years ago the rumblings of an impending volcanic eruption sounded ominously across the Yellowstone country. Suddenly, in a mighty crescendo of deafening explosions, tremendous quantities of hot volcanic ash and pumice spewed from giant cracks at the earth's surface. Towering dust clouds blackened the sky, and vast sheets of volcanic debris spread out rapidly across the countryside in all directions, covering thousands of square miles in a matter of minutes with a blanket of utter

72

devastation. Abruptly, a great smoldering caldera 30 miles across, 45 miles long, and several thousand feet deep—appeared in the central Yellowstone region, the ground having fallen into the huge underground cavern that was left by the earth shaking eruptions. Lava then began oozing from the cracks to fill the still smoking caldera. The third known supervolcano eruption of Yellowstone had occurred. The first two occurred 2 million and 1.2 million years ago. This frequency *suggests* a recurrence rate of one eruption approximately every 600,000 years. When Yellowstone erupts again, it poses not only a general problem with forecasting but also a possible global threatening crisis.

The Scenario

Yellowstone is America's first and most famous National Park. Every year over 3 million tourists visit this stunning wilderness, but beneath its hot springs and lush forests lies a monster of which most of the public is completely unaware.

Most people's idea of a volcano is a symmetrical cone and this involves magma coming up, reaching the surface, being extruded either as lava or as explosive eruptions with ash, and these layers of ash and lava gradually accumulate until you're left with a classic cone shape. We volcanologists know this smooth flowing magma contains huge quantities of volcanic gases, like carbon dioxide and sulfur dioxide. Because this magma is so liquid these gases bubble to the surface, easily escaping. There are thousands of these normal volcanoes throughout the world. Around 40-60 erupt every year, but supervolcanoes are very different in almost every way.

First, they look different. Rather than being volcanic mountains, supervolcanoes form depressions in the ground. Despite never having seen a supervolcano erupt, by studying the surrounding rock scientists have been able to piece together how supervolcanoes are formed. Like normal volcanoes they begin when a column of magma rises from deep within the Earth. Under certain conditions, rather than breaking through the surface, the magma pools and melts the Earth's crust turning the rock itself into more thick magma.

Although it is not clearly understood why, but in the case of supervolcanoes a vast reservoir of molten rock eventually forms. The magma here is so thick and viscous that it traps the volcanic gases building up colossal pressures over thousands of years. When the magma chamber eventually does erupt its blast are hundreds of times more powerful than normal draining the underground reservoir. This causes the roof of this chamber to collapse forming an enormous crater. All supervolcano eruptions form these subsided craters. They are called calderas.

Such is the case with Yellowstone—it is the largest single active system yet discovered. Figure 13-1 below shows the Yellowstone-Teton Geologic System.

Figure 13-1. Yellowstone-Teton Geologic System

There are currently 23 permanent seismographs that are spread across the Park. They detect the sound waves which come from earthquakes deep underground. These waves travel at different speeds depending on the texture of what they pass through. Sound waves passing through solid rock go faster than those traveling through molten rock or magma. By measuring the time they take to reach the seismographs one can tell what they've passed through. Eventually this builds up a picture of what lies beneath Yellowstone. Figure 13-2, below, illustrates the location of the current seismometers around the Yellowstone system.

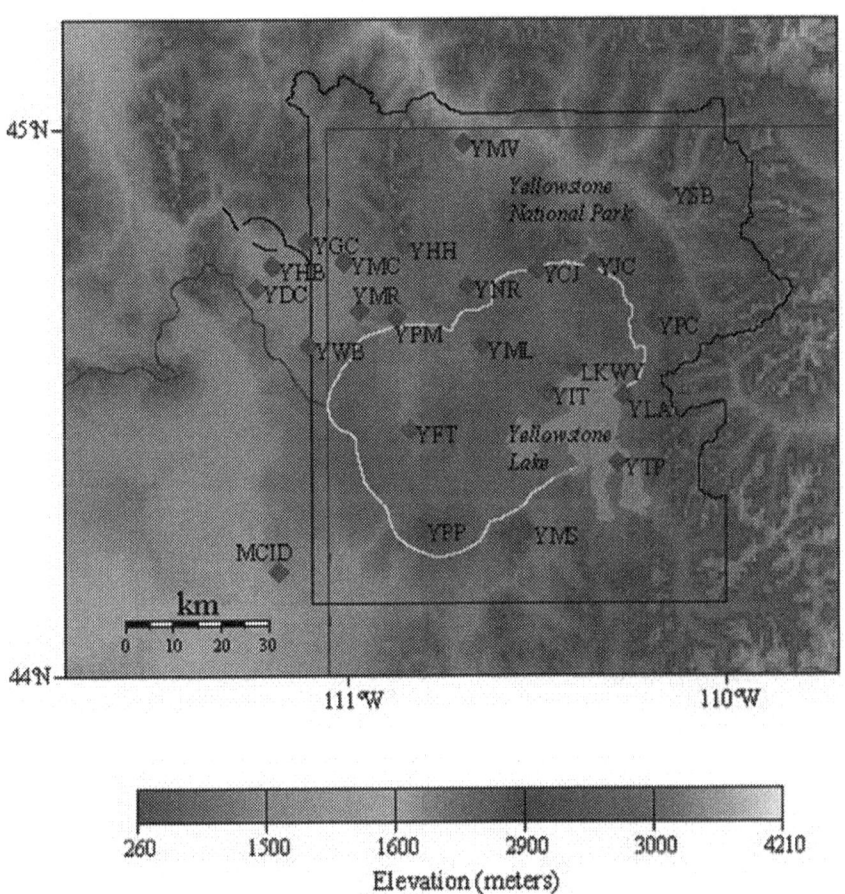

Figure 13-2. The current seismometers around Yellowstone.

The magma chamber that was found extends basically beneath the entire caldera. It is approximately 40-50 kilometres long, approximately 20 kilometres wide and it has a thickness of about 10 kilometres. So it's a giant in volume and essentially encompasses a half or a third of the area beneath Yellowstone National Park.

The Problem

Conventional eruption forecasting of "normal" volcanoes is difficult enough. The forecasting of supervolcanoes presents even more difficult problems. Eruption forecasting has made some progress with the advent of the software package, ***Eruption Pro 10.6***, and its ability to correctly forecast conventional volcanoes eruptions. The current goal of forecasting volcanic eruptions is to provide the best forecasts possible based on the geologic history of the volcano under study as well as on the day-to-day vitals signs of the volcano in terms of earthquakes, surface deformation, temperature, gas emissions, and other measurements. While all of these characteristic vital signs would also apply to supervolcanoes, there are other problems that arise.

The eruption of a conventional volcano is not considered a *rare event* albeit that some volcanoes erupt rarely. In terms of supervolcanoes, eruptions are *extremely* rare events. With respect to rare event statistics, if we consider a rare event, **E**, with a probability of $1/n$ where n is a "large" number, we do not expect the event (eruption) to occur in a single trial (year), i.e., we are surprised if it does. However, in a sequence of trials, the chance of an eruption occurring becomes more likely.

The question at hand becomes, What is the probability of **E** occurring at lease once in n trials? $(1 - 1/n)^n$ is the probability that **E** does not occur at all.

$$\therefore \quad \mathbf{Pr(E \text{ once or more})} = 1 - [\ 1 - (1/n)]^n$$

But if n is large,

$$1 - (1/n)^n \sim e^{-1}$$

$$\therefore \quad Pr(E \text{ once or more}) \cong 1 - e^{-1} = 0.63$$

If one considers one version of Chauvenet's criterion, which is to throw away and event whose probability of occurrence is $\leq \frac{1}{2} n$. The probability of a legitimate occurrence in n trials of an event whose probability is $1/2n$ is

$$1 - [\, 1 - (1/2n)]^n = 1 - \{\, [1 - (1/2n)]^{2n} \,\}^{1/2}$$

$$\cong 1 - e^{-1/2}$$

$$\cong \text{ for } n \text{ large}$$

$$= 1 - 0.606 = 0.39$$

The reader should judge for himself or herself if the straight forward application of the criterion is wise. Probably not! Therefore, we can conclude that conventional rare event statistics **will not** work with supervolcanoes.

Other Techniques

Conventional techniques such as those currently used by SWVRC's *Eruption Pro 10.6*, are also inadequate to properly forecast the eruption of the Yellowstone caldera. The primary reason is that we do not have a history of the volcanoes previous eruptions. Only the current seismicity and deformation parameters can be input to the software programme. This provides an inadequate set of input parameters in order for the programme to properly forecast.

Another probability distribution that may be considered is the negative binomial distribution function. The negative binomial distribution is used when the number of successes is fixed (in this case 3 since there have been three known eruptions to date) and we're interested in the number of failures before reaching the fixed number of successes. In this case, we now suppose that the trials are continued until the event has occurred exactly r times. We want to determine the probability, $\Pr(N = n)$, that this will require **exactly** n trials where N is the number of trials.

$\Pr(N\text{-}n) = \Pr\{[(r-1)$ events in the first trial $(n-1)$ trials] *and* one event on the nth trial)$\} = \Pr\{(r-1)$ events in $(n-1)$ trails$\}\Pr\{1$ on nth$\}$

$$= {}_{(n\text{-}1)}C_{(r\text{-}1)}\ p^{(r\text{-}1)}\ q^{(n\text{-}r)} = b_{neg}(n)$$

$$\therefore \quad b_{neg}(n) = {}_{(n\text{-}1)}C_{(r\text{-}1)}\ p^{(r\text{-}1)}\ q^{(n\text{-}r)}, n = r, r+1,\ldots\ldots$$

This distribution is called the negative binomial since the probabilities may be obtained from successive terms of the expansion of a negative binomial, for example,

$$\sum_{n=r}^{\infty} {}_{(n\text{-}1)}C_{(r\text{-}1)}\ p^r\ q^{n\text{-}r} = p^r\ (1-q)^{\text{-}r}$$

Which may be written

$$= (Q-P)^{\text{-}r}$$

where $Q = 1/p$, $P = q/p$, and $Q - P = (1/p)(1\text{-}q) = 1$

We may rewrite $b_{neg}(n)$ by substituting $s = n - r$ whence

$$\Pr(s) = {}_{(r+s\text{-}1)}C_{(r\text{-}1)}\ p^r\ q^s$$

In the case of Yellowstone, with r = 3, p = 1.67E-06, and x = 640,000 then this calculates out to a .99999967 probability of failure from year to year at this time.

The Consequences

Supervolcanoes are eruptions and explosions of catastrophic proportions. These types of eruptions are absolutely apocalyptic in scale. It is difficult to imagine an eruption this tremendous. The main factor governing the size of this type of eruption of the amount of magma available. If an enormous amount of magma has accumulated in the crust, then you have the potential for a very, very large eruption.

The exact geological conditions needed to create a vast magma chamber exist in only a very few places on earth, so there are only a few known supervolcanoes in the world. The last one to erupt was Toba 74,000 years ago. No modern human has ever witnessed a supervolcano eruption. Volcanologists are not even sure where all the supervolcanoes are but one that is known is Yellowstone National Park.

When Yellowstone goes off again, and it will, it will be a disaster for the United States and eventually, for the whole world. We volcanologists believe it would all begin with the magma chamber becoming unstable. Observations would begin by seeing bigger earthquakes, greater uplifting as magma intrudes and gets nearer and nearer the surface. An earthquake may send a rupture through a brittle layer similar to breaking the lid off a pressure cooker. This would generate sheets of magma, which will perhaps rise up to 30, 40 or 50 kilometres sending gigantic amounts of debris into the atmosphere. Pyroclastic flows would cover the whole region, killing tens of thousands of people in the surrounding area.

The ash carried in the atmosphere and deposited over vast areas of the United States would have devastating effects. A plume of material that goes up into the atmosphere, globally, from the eruption would produce the climatic effects. This would spread worldwide and have a cooling effect that would most likely destroy the growing season on a global scale.

As Dr. Ted Nield, of the Geological Society of London, stated once, *"When a supervolcano goes off, it is an order of magnitude greater than a normal eruption. It produces energy equivalent to an impact with a comet or an asteroid."* *"You can try diverting an asteroid, but there is nothing at all you can do about a supervolcano."*

The eruption will throw out cubic kilometres of rock, ash, dust, sulfur dioxide and so on into the upper atmosphere, where it will reflect incoming solar radiation, forcing down temperatures on the earth's surface. It would be the equivalent of a nuclear winter. The effects would last for four or five years with crops failing and the whole ecosystem breaking down.

Current Monitoring

There are currently 23 seismic monitoring stations about Yellowstone and the site has been examined on a number of occasions for deformation progress/decline.

Yellowstone Park had last been surveyed in the 1920s when the elevation, the height above sea-level, was measured at various points across Yellowstone. 50 years later, Dr. R. B. Smith surveyed the same points. Smith stated *"The idea was to survey their elevations and to compare the elevations in the mid-70s to what they were in 1923 and the type of thing that we did is to make recordings at a precision level of a few millimetres."* The two sets of figures should have been similar, but as the survey team moved across the Park, they noticed something unexpected: the ground seemed to be heaving upwards. The results of the survey indicated that this caldera has uplifted at that time 740 millimetres in the middle of the caldera. As the measuring continued, it became apparent that the ground beneath the north of Yellowstone was bulging up, tilting the rest of the Yellowstone Park downwards (see Figure 13-3 below.) As we venture into the 21st century, once again the Yellowstone caldera appears to be on the uplift swing of the cycle.

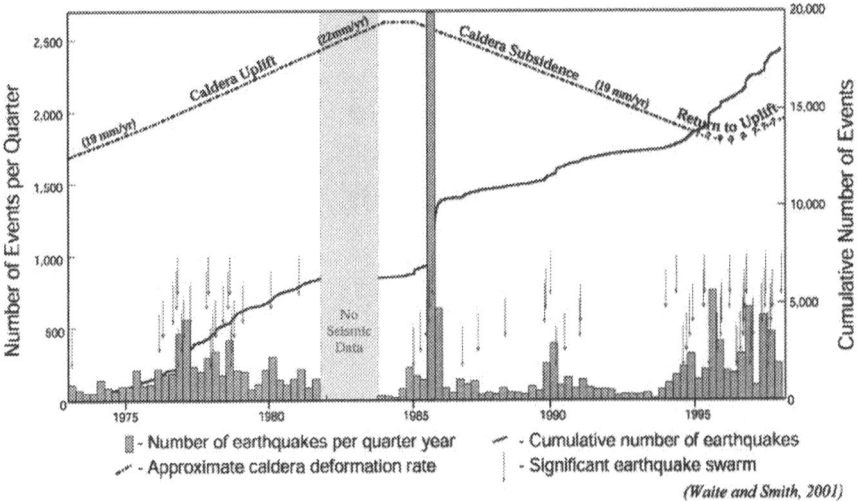

(Waite and Smith, 2001)

Figure 13-3. Caldera uplifting/subsidence and associated earthquake phenomenae.

Things To Look For In Monitoring

We volcanologists believe that the eruption of the Yellowstone caldera would all start with the magma chamber becoming unstable. As previously mentioned, one would probably start seeing bigger earthquakes, also you may see parts of Yellowstone uplifting as magma intrudes and gets nearer and nearer the surface. And maybe an earthquake sends a rupture through the brittle layer, as if you've broken the lid of the pressure cooker. This would generate sheets of magma which, will be probably rising up to 30, 40, 50 kilometres sending gigantic amounts of debris into the atmosphere.

Conclusions:

The only reasonable conclusion that one can come to in studying the current Yellowstone caldera environment is that there is no current way to reasonably and accurately forecast the eruption of the Yellowstone caldera. There has been no eruption of a supervolcano in recent times and although scientists have never witnessed a supervolcano eruption, they can calculate how large they are. There is no recorded data available on the last eruption of Yellowstone, or any other supervolcanoes (e.g., Toba) for that matter. This writer agrees with the observation of colleague Dr. Michael Rampino, *"It's really not a question of if it'll go off, it's a question of when because sooner or later one of these large super eruptions will happen."*

References

1. Meyer, S.L., 1975, *"Data Analysis For Scientists And Engineers"*, John Wiley & Sons, Inc., Pg. 189 & Pg. 283

2. Rampino, M. R., and Self, S., 2000, *"Encyclopedia of Volcanoes, Volcanism And Biotic Extinctions"*, Pgs. 1090-1091

3. Smith, R.B., and Braile, L.W., 1994, *"The Yellowstone hotspot"*: Journal of Volcanology and Geothermal Research, v. 61, pages 121-187.

4. Smith, R.B., and Christiansen, R.L., 1980, *"Yellowstone Park as a window on the Earth's interior"*: Scientific American, volume 242, pages 104-117.

5. Trombley, R. B., & Toutain, J. P., 2002, *"Eruption Pro 10.3—The New & Improved Long-Range Eruption Forecasting Software""*, paper to be presented at the 16th Caribbean Geological Conference, 16 June–21 June 2002, Bridgetown, Barbados.

ABOUT THE AUTHOR

Dr. Robert B. Trombley is an award winning scientist and is Professor Emeritus at the DeVry University in Phoenix where he was an Associate Professor of Mathematics & Computer Sciences. In 1993, he was the visiting volcanologist at the Observatoire Volcanologique du Piton de la Fournaise on France's Réunion Island located in the Indian Ocean completing investigative research on volcano Piton de la Fournaise. Dr. Trombley has visited, studied and researched volcanoes all over the world. He has been an educator and scientist for over 42 years and is a registered volcanologist with the Institut de Physique du Globe de Paris (IPGP) and both a member and contributor to the American Geophysical Union (AGU). He is the winner of the coveted Bauch-Lomb Honorary Science Medal awarded for outstanding contributions in the field of science. In 1988, Dr. Trombley was elected to *"Who's Who In America"*—Midwest Edition, primarily for his scientific contributions.

He has been a consultant to the National Geographic for several TV documentaries on volcanoes. This is his second book involving volcanoes, the first, *"Island Born of Fire: Volcano Piton de la Fournaise"*, was published in 1993 by McGraw-Hill and was re-released as a CD version in 2006.

Dr. Trombley received his undergraduate degree in electronic engineering from the Lawrence Technological University in 1965 and also received an LL.B from LaSalle University in 1973. He obtained his Ph.D. in 1974 from University of Dallas and completed additional post-doctorate training in particle physics fro the University of California Los Angeles. Dr. Trombley has also completed post-doctorate training in geology and paleogeology from the University of New Mexico.

Dr. Trombley has been actively pursuing the science of volcanology for the past forty years, concentrating primarily on the very difficult and controversial task of volcanic eruption forecasting techniques. He has recently developed a new computer software package (*Eruption Pro 10.6*) where he has loaded post-Holocene historic and current eruption data on 499 active volcanoes throughout the world and has performed a statistical analysis of that data. To date, there have been very favourable and accurate forecasts made as a result, particularly in the 50 and 95 percent probability of eruption calculation. Research into volcanic eruption forecasting continues to occupy a considerable amount of scientific research and software development.

APPENDIX-A
GLOSSARY

Aa

A type of lava having a jagged, rough edge and surface.

Active Volcano

Generally volcanoes that have erupted in recorded history or are currently erupting.

Aerosol

A suspension of solid particulates or fine liquid in air.

Andesite

A gray coloured volcanic rock common to stratovolcanoes having a silica content between basalt and dacite.

Ash

In the volcanic sense, fine fragments of lava or rock, about dust size, formed by volcanic explosions.

Ash Cloud

A cloud of ash formed by a volcanic eruption or by a pyroclastic flow.

Avalanche	A large quantity of rocks, volcanic debris, earth, etc., descending swiftly down a mountain side.
Basalt	A type of lava, dark in colour, rich in iron and magnesium and containing approximately 50% silica.
Caldera	A large bowl or basin shaped depression at a volcano's summit generally formed by collapse.
Cinder Cone	A relatively steep mound or hill formed by the accumulation of cinders, usually around a vent of cinders and other volcanic fragments thrown out during an eruption.
Composite Volcano	Same as a Stratovolcano.
Compressional Margin	The edges of convergence of two tectonic plates.
Conduit	A tube or crack through which lava moves.

Continental Crust   The solid outer layer of the Earth beneath the continents. The continental crust is thicker and less dense than the oceanic crust. The continental crust is approximately 25 kilometres in thickness.

Continental Drift   The theory that the slow, relative movements of the continents is caused by horizontal movements of the Earth's surface.

Crater   A bowl-shaped depression around the mouth of a volcano.

Curtain of Fire   A wall of lava fountains that are erupting along a fissure.

Dacite   A light-coloured volcanic rock with intermediate silica composition ranging between rhyolite and andesite.

Dike   A mass of intrusive rock, usually blade-shaped, that cuts layers of surrounding country rock.

Dormant Volcano A volcano that is not currently erupting but is likely to do so in the future.

Earthquake Swarm A sequence of earthquakes, closely spaced in time, of approximately the same magnitude as opposed to a sequence of strong earthquakes with a series of diminishing after-shocks.

Earthquake Wave A vibrational wave produced by an earthquake.

Effusive Eruption An eruption consisting mostly of lava flows as opposed to an explosive eruption.

Eruption Cloud A cloud consisting of gases, ash, and other volcanic fragments which are generated by a volcanic eruption.

Explosive Eruption A sudden expansion of gases ladened with volcanic fragments and other volcanic debris.

Extensional Margin The edges of tectonic plates that are moving apart from one another.

Extinct Volcano A volcano that has not erupted in historic times or is not expected to erupt again in the future.

Fault A crack or fracture in the Earth's crust along which there has been evidence of movement.

Fault Scarp A cliff formed along a fault caused by movement of the fault.

Feldspar A light-coloured mineral composed principally of oxygen, aluminum, and silicon.

Felsic Crust Crust that is made up of a light coloured igneous rock that is poor in iron and magnesium and contains abundant feldspars and quartz.

Fissure A large blade-shaped fracture in the Earth's crust.

Flank Eruption	An eruption that vents from the side of a volcano instead of from the summit of the volcano.
Fumarole	An opening (vent) in the surrounding earth from which volcanic gases and steam are emitted.
Geophysics	The study of the physical and mechanical aspects of the science of geology.
Geothermal Energy	The energy which is derived from the Earth's internal heat source.
Geothermal Gradient	The rate of change in temperature with depth in the Earth.
Geothermal Power	The power generated by the heat energy of the Earth.
Granite	A course-grained igneous rock composed chiefly of quartz and feldspar.
Guyot	A flat-topped submarine mountain or seamount.

Harmonic Tremor   A volcanic tremor that has a relatively steady state frequency and amplitude.

Holocene   The period of geologic time since the last major ice age; approximately 10,000 years ago to the present day.

Hot-Spot Volcano   A volcano related to a persistent heat source in the mantle.

Hydrothermal   An area that lies underground containing porous rock that contains a reservoir of hot water.

Igneous Rock   Magma or lava that has cooled and solidified either below or above the Earth's surface.

Ignimbrite   A widespread deposit that has been left by a large pyroclastic flow.

Intrusion   A body of rock formed by magma that has forced its way into the surrounding host rock and then cools. It is also the process of forming such a body of rock.

Island Arc

A curving chain of volcanic islands, such as the Hawaiian chain, formed at compressional plate margins.

Lahar

A mudflow created by volcanic activity such as the melting of ice and snow.

Lava

Magma that has reached the surface of the Earth. It is also what the resulting rock is called after cooling.

Lava Channel

A swift-moving, incandescent portion of an active lava flow or its solidified remains.

Lava Dome

A mass of viscous lava, steep-sided, usually with a rounded shaped top which covers a volcanic vent.

Lava Flow

A stream of molten rock, effusive in nature that moves down the side of a volcano.

Lava Fountain

A very rapid jet of incandescent lava sprayed from a volcanic vent caused by the rapid expansion of gases.

Lava Lake

Literally, a lake of molten lava within a volcanic crater or depression.

Lava Tube

A tunnel beneath the surface of a solidified lava flow. Also the cave formed by the emptying of a tunnel as the eruption ceases or shifts direction.

Magma

Molten rock within the Earth. Magma that reaches the surface of the Earth is refereed to as lava.

Magma Chamber

The underground reservoir in which magma is stored.

Magmatic Gases

Various gases such as carbon dioxide, hydrogen sulfide and water that is contained (dissolved) in magma.

Mantle	The zone of the Earth below the crust to a depth of approximately 3500 kilometres. Also that area that is located above the core of the Earth.
Microearthquake	An earthquake that is too small to be felt at the surface of the Earth but is detectable with a seismograph.
Mudflow	A water-saturated mixture of mud and debris that flows downslope under the force of gravity. Also known as a Lahar.
Neck	A vertical intrusion, usually seen as an erosional remnant, that depicts a former volcanic conduit.
Normal Fault	An inclined fault whereby the upper block moves downward relative to the lower block.
Nuée Ardente	(Fr) A dense "glowing cloud" of volcanic gas and ash that erupts from a volcano and moves swiftly down the slopes as a pyroclastic flow.

Obsidian	A dark coloured or black volcanic glass that is generally composed of rhyolite.
Oceanic Crust	The crust of the Earth where it lies under the oceans and without the layer of granite that forms the continents. It is generally approximately 5 kilometres in thickness.
Oceanic Ridge	A major mountain range that lies completely underwater.
Olivine	An olive-green mineral composed of silicon, iron, magnesium and oxygen.
Ore	Any rock material from which minerals of commercial value can be extracted.
Pahoehoe	A basaltic lava with a smooth "ropy" or billowy textured surface.
Partial	A stage of cooling magma where it is composed of partly solid crystallization crystals and partly liquid rock.

Partial Melt A stage of melting rock when it is partly solid and partly liquid crystals.

Pelee's Hair Natural spun glass in strands formed generally in lava fountains.

Pillow Lava Rounded, bag or sac like bodies of lava that form underwater as it is extruded.

Plate Tectonics The theory that the Earth's crust is broken into many pieces, about a dozen or so plates, that slowly move about in relation to one another.

Plume A column of hot, plastic rock rising from well within the mantle to form what are known as "hot-spot" volcanoes.

Pluton A large igneous intrusion that cools and solidifies beneath the surface of the Earth.

Pumice
A form of volcanic glass filled with so much gas bubbles and holes that it resembles a sponge and is very light in weight.

Pyroclastic Deposit
The deposit of volcanic fragments from a pyroclastic flow.

Pyroclastic Flow
Same as a nuée ardente.

Quartz
A rock forming mineral composed chiefly of silicon and oxygen.

Repose Time
The time interval between eruptions.

Rift System
The oceanic ridges, greater than 60,000 kilometres in length, where the tectonic plates are separating and new crust is being formed. An example of the surface counterpart is the East African Ridge.

Rift Volcano
A volcano along a rift system.

Rift Zone
A region whereby the crust is separating and pulling apart.

Ring of Fire

The region of converging tectonic plate margins, with the resulting earthquakes and volcanoes, that surrounds the Pacific Ocean.

Rhyolite

A fine-grained volcanic rock, similar in nature to granite, but with a high silica composition.

Seafloor Spreading

The creation of new seafloor at the oceanic ridges as the tectonic plates separate from one another.

Seamount

A mountain, usually alone and volcanic, that lies underwater.

Seismic Wave

Same as an earthquake wave.

Seismograph

An instrument used for recording seismic waves in the crust of the Earth.

Seismology

The study of seismic waves, earthquakes, and the interior of the Earth's structure.

Shield Volcano   A volcano which is built by successive flows of fluid basaltic lava and forms a dome shaped structure with gentle sloping sides.

Silica   A chemical composition comprised of silicon and oxygen.

Silicate Mineral   A mineral composed mainly of silicon and oxygen.

Silcic   A term for volcanic rock or magma that is rich in silica and is similar to rhyolite.

Solfatara   A fumarole that has gases that are principally sulfurous.

Spatter Cone   A cone built up around a vent by fragments of still molten lava that weld into a large mass.

Stratovolcano   A steep sided volcanic cone built by a combination of lava flows and pyroclastic flows from explosive eruptions.

Strike-Slip Fault

A predominantly vertical fault with sideslipping, horizontal displacement.

Subduction-Type

A volcano that occurs just inland from a subduction zone such as those volcanoes along the Cascade range.

Subduction Zone

The zone where two tectonic plates converge generally with one plate overriding the other plate.

Surge

A short timed increase in the velocity and volume of a lava flow.

Talus

A gathering of rock debris at the base of a cliff or steep slope.

Tephra

Material of all sizes and types that erupts from a volcano and is usually deposited by airfall. Also another name for pyroclastic deposits.

Thermal Gradient   The rate of change of temperature with increase in depth or distance.

Thrust Fault   A gently inclined fault whose upper side moves relatively upward.

Tidal Wave   Same as a tsunami.

Transform Fault   A strike-slip fault that connects the offsets of a mid-oceanic ridge.

Tsunami   A giant sea wave produced by an underwater earthquake, landslide or volcanic eruption.

Vein   An opening in the surface of the Earth through which volcanic material are erupted.

Viscosity   A measure of the resistance to flow in a liquid.

Volcanic Block   Volcanic debris fragments, generally larger than 64mm in size, that are thrown out during an explosive eruption.

Volcanic Bomb A lump of molten lava which is thrown out of a volcano and takes on a rounded shape.

Volcanic Cinder A lava fragment approximately 1 cm in diameter.

Volcanic Complex A persistent volcanic event area that has, over a period of time, built a complicated mixture of volcanic landforms.

Volcanic Dust Fine particulates of volcanic ash.

Volcanic Front A line of volcanoes that are closest to an oceanic trench.

Volcanic Tremor A continuous vibration of the ground, detectable by seismograph that is associated with volcanic eruptions and other subsurface volcanic activity.

Welded Tuff Pyroclastic deposits, so hot when formed, that the fragments weld together forming a solid rock.

APPENDIX-B
LUNAR CALCULATIONS

$$a_e = \frac{G \, M_m}{d^2}$$

$G = 6.67\text{E-}11 \text{ N m}^2 \text{ kg}^{-2}$

$M_m = 7.35\text{E+}22 \text{ kg}$

$d = 384{,}400 \text{ km}$

Therefore, $a_e = 3.32\text{E+}01$

$$f_A \approx f_B \approx 2 \, \frac{M_m}{M_e} \left[\frac{\alpha}{d}\right]^3 G$$

$M_m = 7.35\text{E+}22 \text{ kg}$

$M_e = 5.98\text{E+}24 \text{ kg}$

$a_e = 3.32\text{E+}01$

$G = 6.67\text{E-}11 \text{ N m}^2 \text{ kg}^{-2}$

$d = 384{,}400 \text{ km}$

Therefore, $f_A \approx f_B \approx 1.06\text{E-}24$

APPENDIX-C
SOLAR CALCULATIONS

Distance of Earth to Moon =   3.844E+05 km

Distance of Earth to Sun =   1.496E+08 km

Distance to Earth-Moon Centre =   1.922E+05 km

Distance to Earth-Moon Centre to Sun =   1.494E+08 km

Earth Mass =   5.980E+24 kg

Moon Mass =   7.349E+22 kg

Sun Mass =   1.973E+30 kg

$G = \quad 6.672\text{E-}11 \text{ N m}^2 \text{ kg}^{-2}$

$$a_e = G * \frac{M_s}{D_s^{\ 2}} \quad = \quad 6.672\text{E-}11 * \frac{1.973\text{E+}30}{(1.494\text{E+}08)^2}$$

$$a_e \quad = \quad 5.90\text{E+}03$$

$$E_s = \frac{M_s}{M_e} \left[\frac{a_e}{d}\right]^3 = \frac{1.973\text{E+}30}{5.980\text{E+}24} * \left[\frac{5.980\text{E+}03}{1.494\text{E+}08}\right]^3$$

$$E_s \quad = \quad 2.02\text{E-}08$$

APPENDIX-D
ERUPTION PRO 10.6'S OUTPUT—MEANINGS & ACRONYMS

Avg. Repose

The average number of years of the repose interval for this volcano.

C/L Interval

The Confidence Level interval a statistical range used in the repose analysis.

Location

The country where the volcano is located followed by the latitude and longitude.

LUNAR

Probability contribution attributed to lunar influences.

Elevation

The height of the volcano expressed in metres and feet.

Forecasted Year Limit

Year that the volcano should erupt with respect to repose only.

Microgravity Monitored   This volcano is monitored for gravity changes thru instrumentation (Yes/No).

Number of Eruptions   The number of recorded eruptions in the volcano's entire history.

PDEF   Probability contribution attributed to deformation of the volcano.

Projected Yrs. Max.
Until Eruption   Number of years from the current year before an eruption event is expected with respect to repose only.

PSO2   Probability contribution attributed to sulphur dioxide (SO_2) measure-ments.

PTHERM   Probability contribution attributed to real-time thermal imaging of volcanoes.

Pr(0)   Probability of no eruption for the current year.

Pr(1)   Probability of an eruption event
                             for the current year.

PSI   Probability contribution attributed
                             to seismic activity.

SOLAR   Probability contribution attributed
                             to solar influences.

SSPOT   Probability contribution attributed
                             to sunspot influences.

Statistically Forecasted
Eruption Year   Year that the software forecasts the
                             volcano to have an eruptive event.

Thermal Imaging
Available   This volcano is monitored for ther-
                             mal changes thru satellite instru-
                             mentation (Yes/No).

Total Years   Number of years from the very
                             first eruption of the volcano to the
                             current year.

U   The calculated eruption rate/year
                             used in the Poisson distribution
                             calculation.

VEI	Volcano Explosivity Index on a scale of 0 to 8. (See Appendix-E for full details of VEI).
VFEA	Volcano Frequency of Eruption Analysis—a probability contribution (0-1) based on the eruption history of the volcano for the last 10 years.
Volcano	The volcano's name followed by the type of volcano it is.
Year of 1st Eruption	The year that the first recorded of the volcano occurred.
Year of Forecasted Eruption @ ≥50%	The starting year that the software forecasts The volcano to have an eruptive event with ≥50% probability.
Year of Forecasted Eruption @ ≥95%	The starting year that the software forecasts The volcano to have an eruptive event with ≥95% probability.
Year of Last Eruption	The year that the volcano last erupted.

APPENDIX E
VOLCANIC EXPLOSIVE INDEX
(VEI) CHART

VEI	Description	Volume of Ejecta(m^3)	Column Height (Km)	Qualitative Description	Classification
0	Non-Explosive	$<10^4$	<0.1	Gentle Effusive	Hawaiian
1	Small	10^4-10^6	0.1 - 1	Gentle Effusive	Hawaiian-Strombolian
2	Moderate	10^6-10^7	1 - 5	Explosive	Strombolian
3	Mod-Large	10^7-10^8	3 - 15	Explosive	Strombol-Vulcanian
4	Large	10^8-10^9	10 - 25	Cataclysmic	Plinian-Vulcanian
5	Very Large	10^9-10^{10}	>25	Cataclysmic	Vulcanian-Plinian
6	Very Large	10^{10}-10^{11}	>25	Paroxysmal	Plinian-Ultra-Plinian
7	Very Large	10^{11}-10^{12}	>25	Colossal	Plinian-Ultra-Plinian
8	Very Large	$>10^{12}$	>25	Colossal	Ultra-Plinian

978-0-595-41260-0
0-595-41260-2

www.ingramcontent.com/pod-product-compliance
Lightning Source LLC
Chambersburg PA
CBHW051433280526
45785CB00003B/1276